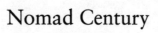

Nomad Century

Nomad Century

*How Climate Migration Will
Reshape Our World*

GAIA VINCE

FLATIRON
BOOKS
NEW YORK

www.flatironbooks.com

Library of Congress Control Number: 2022941099

ISBN 978-1-250-82161-4 (hardcover)
ISBN 978-1-250-84711-9 (ebook)

Our books may be purchased in bulk for promotional, educational, or business use. Please contact your local bookseller or the Macmillan Corporate and Premium Sales Department at 1-800-221-7945, extension 5442, or by email at MacmillanSpecialMarkets@macmillan.com.

First published in 2022 by Allen Lane, part of the Penguin Random House group of companies

First U.S. Edition: 2022

10 9 8 7 6 5 4 3 2 1

*For my father
And for all those who nurture tropical
flowers under grey northern skies*

Contents

Contents

List of Figures

Introduction

A great upheaval is coming. It will change us, and our planet.

In the global south, extreme climate change will push vast numbers of people from their homes, with large regions becoming uninhabitable; in the planet's more comfortable north, economies will struggle to survive demographic changes with massive workforce shortages and an impoverished elderly population.

Over the next fifty years, hotter temperatures combined with more intense humidity are set to make large swathes of the globe lethal for 3.5 billion of us. Fleeing the tropics, the coasts and formerly arable lands, huge populations will need to seek new homes; you will be among them, or you will be receiving them. This migration has already begun – we have all seen the streams of people fleeing drought-hit areas in Latin America, Africa and Asia where farming and other rural livelihoods have become impossible. Climate-driven movements are adding to a massive migration already under way to the world's cities. The number of migrants has doubled globally over the past decade, and the issue of what to do about rapidly increasing populations of displaced people will only become greater and more urgent as the planet heats.

Have no doubt, we are facing a species emergency – but we *can* manage it. We can survive, but to do so will require a planned and deliberate migration of a kind humanity has never before undertaken.

People are finally beginning to face up to the climate emergency. However, while nations rally to reduce their carbon emissions, and try to adapt at-risk places to hotter conditions, there is an elephant in the room: for large portions of the world, local conditions are becoming too extreme and there is *no way to adapt*. The world already sees twice as many days where temperatures exceed 50°C than thirty years ago – this level of heat is deadly for humans, and also hugely problematic for buildings, roads and power stations. In short, it makes an area unliveable.

This explosive planetary drama demands a dynamic human response, and the solutions are within our hands. We need to help people to move from danger and poverty to safety and comfort – to build a more resilient global society for everyone's benefit. Human movement on a scale never before seen will dominate this century and remake our world. It could be a catastrophe or, managed well, it could be our salvation.

People will have to move to survive.

Large populations will need to migrate, and not simply to the nearest city, but also across continents. Those living in regions with more tolerable conditions, especially nations in northern latitudes, will need to accommodate millions of migrants into increasingly crowded cities while themselves adapting to the demands of climate change. We will need to create entirely new cities near the planet's cooler poles, in land that is rapidly becoming ice-free. Parts of Siberia, for example, are already experiencing temperatures of 30°C for months at a time.

Wherever you live now, this migration will affect you and the lives of your children. It may seem obvious that Bangladesh, a country where one-third of the population lives along a sinking, low-lying coast, is becoming uninhabitable. (More than 13 million Bangladeshis – nearly 10 per cent of the population – are expected to have left the country by 2050.) Or that desert nations like Sudan are becoming unliveable. But in the coming decades wealthy nations will be severely affected too. Hot, drought-afflicted Australia will suffer, as will parts of the United States, forcing millions from cities such as Miami and New Orleans to seek safety in cooler states like Oregon and Montana. Cities will need to be built to house them.

In India alone, close to a billion people will be at risk. Another half billion will need to move within China, and millions more across Latin America and Africa. Southern Europe's treasured Mediterranean climate has already shifted north, leaving regular desert-like conditions from Spain to Turkey. Meanwhile, parts of the Middle East have already been made intolerable by increasing heat, lack of water and poor soils.

People will begin leaving. They are already on the move.

*

We are undergoing a species-wide planetary upheaval and it occurs not only at a time of unprecedented climate change but also of human demographic change.

Global population will continue to rise in the coming decades, peaking at perhaps 10 billion in the 2060s. Most of this increase will be in the tropical regions that are worst hit by climate catastrophe, causing people there to flee northwards. The global north faces the opposite problem – a 'top-heavy' demographic crisis, in which a large elderly population is supported by a too-small workforce. At least twenty-three nations, including Spain and Japan, are expected to see their populations halve by 2100. North America and Europe have 300 million people above the traditional retirement age (65+), and by 2050 the economic old-age dependency ratio there is projected to be at forty-three elderly persons per 100 working persons aged 20–64.[1] Cities from Munich to Buffalo will begin competing with each other to *attract* migrants. This competition will become especially acute towards the end of the century, when some of the southern places made uninhabitable by climate change may become once again liveable through geoengineering innovations that reduce global or regional temperatures, through carbon dioxide removal and technological interventions that can cool large areas cheaply. Truly, this is the century of unprecedented, planetary human movement.

We need to plan pragmatically now, adopting a species-wide approach to ensure our human systems and communities have the resilience to weather the shocks to come. We already know which communities will need to relocate by 2050, when I will be in my seventies. We know also which places will be safest at the end of the century, when my children will be in their old age.

We need to look *now* at where these billions of people could be sustainably housed. Doing so will require international diplomacy, negotiations over borders, and adaptation of existing cities. The Arctic, for instance, will become a relatively habitable destination for millions of people, although the current infrastructure there, minimal though it is, is already sinking into the melting permafrost and will have to be rebuilt for the hotter conditions. Preparing for this climate migration means the phased abandonment of major cities, the relocation of others, and the building of entirely new cities in foreign lands.

London, the city I live in, is at least 2,000 years old and accommodates 9 million people. We have mere decades to adapt, expand and build such cities. We can build emergency hospitals in a few days, as we saw during the Covid-19 pandemic; I've no doubt we can build ambitious cities within years. But what sort of cities, and where, and for whom?

The coming migration will be big and diverse. It will involve the world's poorest fleeing deadly heatwaves and failed crops. It will also include the educated, the middle class, people who can no longer live where they planned because it's impossible to get a mortgage or property insurance; because employment has moved elsewhere; because the neighbourhood has become undesirable because those who could have already left for a more tolerable climate. Climate change has already uprooted millions in the US – in 2018, 1.2 million were displaced by extreme conditions; by 2020, the annual toll had risen to 1.7 million people. The US now averages a billion-dollar disaster every eighteen days.[2] A 2021 survey of Americans who were moving home found that half cited climate risks as a factor.

As I write this, more than half of the western US is facing extreme drought conditions, and farmers in Oregon's Klamath Basin are talking about illegally using force to open dam gates for irrigation. At the other extreme, by 2050 half a million existing US homes will be on land that floods at least once a year, according to data from Climate Central, a partnership of scientists and journalists. Those homes are valued at $241 billion. Even if a building doesn't itself flood, if enough local infrastructure is flooded, the neighbourhood becomes unviable and people move away. Those affected will include residents of important cities, such as the 400,000 inhabitants of New Orleans. Louisiana's Isle de Jean Charles has already been allocated $48 million of federal tax dollars to move the entire community due to coastal erosion and rising sea levels. In Britain, the Welsh villagers of Fairbourne have been told their homes should be abandoned to the encroaching sea as the entire village is to be 'decommissioned' in 2045. Larger coastal cities are at risk, too. Consider that the Welsh capital, Cardiff, is projected to be two-thirds under water by 2050.

For you, the coming upheaval may be a sudden, urgent exodus because climate change has devastated harvests, food prices have soared, and your country has been overtaken by violent conflict and become unsafe. Or it may be that a hurricane devastates your town, or ocean waves erode your village. The upheaval will happen suddenly, in the wake of catastrophes, and it will happen slowly, in dribs and drabs. The United Nations International Organization for Migration estimates that there could be as many as 1.5 billion environmental migrants in the next thirty years alone. After 2050, that figure is expected to soar as the world heats further and the global population rises to its predicted peak in the mid 2060s. Disasters already displace up to ten times more people than conflict and war worldwide.

We are making a new and very different world through our environmental changes. As the only sentient beings capable of such audacious planetary transformation, we must have the maturity and wisdom to direct our talents towards saving ourselves.

I've certainly panic-Googled land prices in Canada and New Zealand, seeking a safe place for my children's future with reliable fresh water and greenery for the coming decades. But I have also had to accept that this is not a challenge that we can meet as individuals. For if we approach the greatest migration in a piecemeal way – in which those who can, buy safety in the least affected parts of the world – we risk an inequality of survival that threatens us all. We would face the likelihood of an enormous loss of life, of terrible wars and misery, as the wealthy erect barriers against the poorest. We see this devastating situation occurring in a far smaller way today – we cannot allow such calamitous chaos at the scale expected in a few decades. Quite apart from the moral abhorrence, there would be no peace for any of us. Instead, we must come together as a global society to address this human-made problem. We are a planetary species, dependent on a single shared biosphere. We must look afresh at our world and consider where best to put its human population and meet all of our needs for a sustainable future.

Doing so requires a radical rethink. The question for humanity becomes: what does a sustainable Promised Land look like? If we manage to achieve a commonwealth of humanity, we will continue to dominate the globe, although we and our food production will inevitably be limited to a relatively small region. We will need to develop an entirely new way of feeding, fuelling and maintaining our lifestyles in this Anthropocene era, while also reducing atmospheric carbon levels. We will need to live in denser concentrations in fewer cities, while reducing the associated risks of crowded populations, including power outages, sanitation problems, overheating, pollution and infectious disease.

At least as challenging, though, will be the task of overcoming a geopolitical mindset, the idea that we belong to a particular land and that it belongs to us. In other words, we will, as refugees of nations, need collectively to transition to a sense of ourselves as citizens of Earth. We will need to shed some of our tribal identities to embrace a pan-species identity. We will need to assimilate into globally diverse societies, living in new, polar cities. We will need to be ready to move again when needed.

With every degree of temperature increase, roughly a billion people will be pushed outside the zone in which humans have lived for thousands of years. We are running out of time to manage the coming upheaval before it becomes overwhelming and deadly. Migration is not the problem; it is the solution.

Migration will save us, because it is migration that made us who we are.

I'll begin by showing you the nomadic soul that perches inside us all. Migration is a valid and essential part of our species' nature. Hundreds of thousands of years ago, our ancestors developed the adaptability to live anywhere. It made us the planetary primate.

Even more unusually, humans don't just relocate themselves but we also migrate the stuff of the planet – other animals, plants, water and materials. We rely on creating networks, exchanging our genes, ideas and resources to thrive. Eventually, these networks became so strong that we didn't need to move ourselves, for we could instead summon the bits of the planet we need: a virtual migration. Unlike any other animal, we survive not on the stuff of our physical location but on these virtual migrations we all make continually. I type this paragraph now using components dug from Congolese rock, wearing clothes made in Vietnam, having lunched on potatoes grown in Peru. Human ecology is planetary. It is reconfiguring Earth.

Over the coming decades we face multiple crises including heat and fires, floods and sea-level rise, extreme weather, and demographic shifts in our growing populations. Underlying every one of these, and turning them from hazards into full-blown humanitarian crises, is social inequality and poverty. Climate change is often described as a threat multiplier – the people most affected are those already experiencing threats to their lives and livelihoods, including degraded environments, income instability, inability to save money or resources, lack of affordable healthcare, inadequate sanitation, poor governance, and a lack of personal agency or ability to change their circumstances. The shocks and stresses of climate change hit people with the least resilience hardest, pushing them beyond their ability to cope. We are facing a climate apartheid.

In these chapters, we will explore what some of the emerging crises mean for our world and human populations – and to warn you: it's not good. But stay strong, because then we will see that the solutions are already within reach.

This book looks at where it will be safe to live, how and in what numbers. It will look at where food, power, water and other resources can be produced. Even for those people who are receiving migrants, rather than migrating themselves, life will be an upheaval. Cities will have to be repurposed and adapted to the changing environmental conditions and a vastly swollen population in ways that will render them unrecognizable – but seizing the opportunity to become better. The ways in which we all see and understand each other, as citizens, traders and members of a global society, will be transformed by this new world.

How we manage this global process, and how humanely we treat each other as we migrate, will be key to whether this century of upheaval proceeds smoothly or with violent conflict and unnecessary deaths. Managed right, this upheaval could lead to a new global commonwealth of humanity.

Humans evolved to cooperate, and they also evolved to migrate. The upheaval that awaits us may be unprecedented, but it arises out of a long history founded on this same adaptive behaviour. Now is the time for us to restore this inherent flexibility about where we live.

This a chance to recognize the dependence of all of us on each other, and our species' dependency on the natural world, as we restore its healthy function for the protection of all of us. The final part of the book looks at restoring our planet's habitability so that large human populations can once again live in the tropics. This means reducing the dangerous global temperatures that will characterize this century – something that can be achieved through decarbonizing our energy systems, removing carbon from the atmosphere and by reflecting the sun's heat back to space. I'll look at the latest technological innovations, and the enormous political, social and diplomatic tussles we'll need to reconcile if we are to create a just world for 9 billion of us. As you read this book, I ask you to approach its ideas with an open mind, whichever side of an ideological divide you sit on: quell the impulse to immediately reject radical social solutions as 'implausible' or 'impractical', or technological solutions as 'unnatural' or 'dangerous'. We are social, technological apes – we solve our problems using our exceptional skill in both areas and this, the biggest crisis humanity has faced in the history of our species, will require our holistic toolbox. Neither large-scale technological change nor fundamental social change are easy or comfortable options and both come with significant challenges, but the state that we are in leaves us with few choices. This book is my assessment of our best way forward.

Migration stories moulded my childhood and I have always been drawn to people from other places. As the daughter and granddaughter of refugees and migrants, I have lived on three continents and have travelled widely. On my longest trip, a two-and-a-half-year journey through fifty countries to research my first book, I spoke to princes, presidents and paupers about what it means to lose your home – amongst them, the presidents of the Maldives and of Kiribati, who face tough decisions as their lands disappear with climate change. I have visited the stateless 'char people' who live on the ephemeral mud islets that briefly appear in the Ganges river between India and Bangladesh. And I have spent time living with African and Central American hunter-gatherers, for whom home is never a settled address. Over the past decade, I have been investigating the science of our increasing environmental changes, from the hotter atmosphere to the loss of biodiversity and the growth of farmlands, as we enter the Anthropocene, a world beyond anything experienced in the history of humanity. I have written about the threats and dangers to wildlife and human life, and have made radio and television programmes about how we can adapt to this new world. Yet the most important adaptation for many millions of people – increasingly their only option – is rarely mentioned and seldom advocated: migration.

As a scientist by training, I know that many of the climate changes we face are fixed in place for decades, if not for centuries. The temperature of the planet is already rising, yet even so we continue emitting carbon dioxide. The window for action is closing.

The world: 4°C hotter*

Conditions across a broad tropical belt make large regions uninhabitable, and rising sea levels drown many islands and coastal settlements. However, renewable energy production is possible throughout the globe, and enough food can be produced to feed 10 billion people.

Arctic passage
With no sea ice, this valuable shipping route is open all year, providing transportation links between habitable zones in Canada and Russia

Siberia
Reliable precipitation and warmer temperatures provide ideal growing conditions for crops

Canada
Reliable precipitation and warmer temperatures provide ideal growing conditions in northern areas, although drought will be an issue in central and southern parts

Greenland
Ice sheet will be melting rapidly, exposing new areas for habitation, agriculture and mining

Scandinavia/UK/Northern Russia/Greenland:
Compact high-rise cities could provide shelter for much of the world's population

North Africa/Middle East/Southern US
Solar-Wind Energy Belt stretches for thousands of kilometres, deploying a mixture of wind, photovoltaic and solar thermal energy. At frequent intervals a high voltage direct-current substation sends power north

Southern Europe
Deserts have encroached on the continent, rivers have dried up and the mountains are snow-free

Eastern China
Dried rivers and aquifers mean this region has been abandoned. Intense monsoons have helped erode the land, leaving a dustbowl

South-west US
Desertification, fire and heat will make large portions uninhabitable. The Colorado River is a mere trickle. The land is used for solar farming and geothermal energy

Africa
Mostly desert, though some models show greening of the Sahel

Amazon
Desert

Polynesia
Vanished beneath the sea

New Zealand
Unrecognisable. This densely populated island state has high-rise cities and intensive farming

Patagonia
The southern hemisphere's greatest potential for agriculture and habitation

Australia
In the far north and Tasmania, compact cities house people and crops are grown. The rest of the continent is given to solar energy production, hydrogen and minerals, such as uranium mining for nuclear power

Peru
Deglaciation means this area is dry and uninhabitable

Tropical peaks
Most of the world's highest mountains, from the Himalayas to the Andes, have lost their glaciers, with impacts on major rivers in their regions

Western Antarctica
Ice free with agricultural potential, and perhaps even cities

Legend:
- Food-growing zones/ Compact high-rise cities
- Desertification
- Uninhabitable due to floods, drought or extreme weather
- Potential for reforestation
- Land lost due to rising sea levels, assuming a 2-metre rise
- ::: Solar energy
- Geothermal energy
- ✿ Wind energy

(*than the preindustrial average)

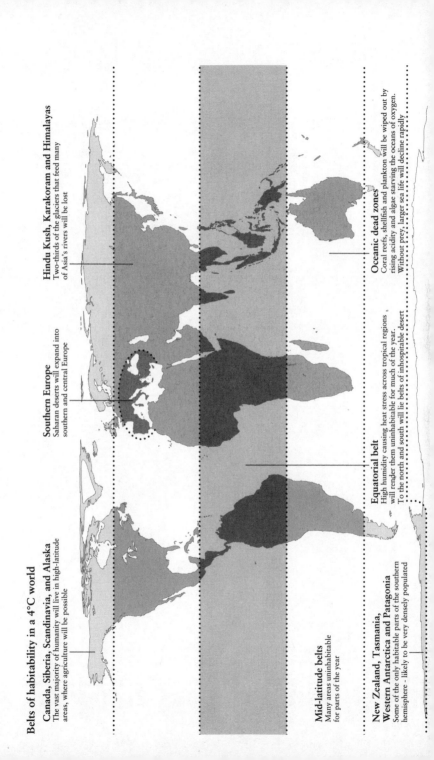

Belts of habitability in a 4°C world

Canada, Siberia, Scandinavia, and Alaska
The vast majority of humanity will live in high-latitude areas, where agriculture will be possible

Southern Europe
Saharan deserts will expand into southern and central Europe

Hindu Kush, Karakoram and Himalayas
Two-thirds of the glaciers that feed many of Asia's rivers will be lost

Oceanic dead zones
Coral reefs, shellfish and plankton will be wiped out by rising acidity and algae starving the oceans of oxygen. Without prey, larger sea life will decline rapidly

Mid-latitude belts
Many areas uninhabitable for parts of the year

Equatorial belt
High humidity causing heat stress across tropical regions, will render them uninhabitable for much of the year. To the north and south will lie belts of inhospitable desert

New Zealand, Tasmania, Western Antarctica and Patagonia
Some of the only habitable parts of the southern hemisphere - likely to be very densely populated

I

The Storm

The forecast is terrible. We face environmental, social and demographic catastrophe: drowned cities; stagnant seas; a crash in biodiversity; intolerable heatwaves; entire countries becoming uninhabitable; widespread hunger; and a global population of some 10 billion humans. A 3–4°C-hotter world is the stuff of nightmares and yet that's where we're headed within decades.

Our problems are systemic and they are feeding into each other to create a catastrophic avalanche for humanity. Polls show that most people around the world are now convinced that we are facing a 'climate emergency',[1] but even this alarming phrase doesn't encompass the sheer scale of disaster, which could be nothing short of global societal breakdown.

The amount of carbon dioxide in the atmosphere, which in 2022 reached 420 parts per million, is already higher than it's been for at least the past 3 million years.[2] It's heating the planet beyond anything humans have experienced during our entire evolutionary history – and fast. As far as we know, only the instantaneous Cretaceous–Palaeogene meteorite impact event, 66 million years ago, caused more rapid global climate change than the current human-induced global heating. During that event, which famously killed off the dinosaurs, about 600–1,000 gigatonnes of carbon dioxide was released (with enormous amounts of other climate-changing gases).[3] Now, *we* are the asteroid, taking just twenty years to release 600 gigatonnes of carbon dioxide.

We have created for ourselves a similarly perilous planetary situation, and we are barely better prepared than the dinosaurs for impending disaster. Collectively, the world has thus far failed to respond to the triple crises of poverty, climate change and ecosystem collapse at the scale and speed so desperately needed by the most vulnerable people.

Take climate change: we know that our carbon dioxide emissions are raising the temperature of the atmosphere and oceans, producing extreme weather events, a rise in sea levels, and altering rainfall patterns across the world. We know this is dangerous, and that we need to stop producing these emissions at a much faster rate – we need not just to match the rate at which carbon dioxide can be removed from the atmosphere, but cut below it. In other words, we need to go beyond 'net zero' emissions and start reducing down to safe levels the amount of carbon dioxide that's already there. We know all this, but the vast, complex human economic, cultural and technological system – of which we are each a part – is slow to shift. We are continuing to chart a path towards a 4°C rise this century.[4]

The big reason for this heating is that global energy use is increasing (and will continue to do so for many decades), and most of this energy is generated by adding carbon dioxide to the atmosphere because it comes from burning fossil fuels. The obvious options then, determined by the physics of planetary heating, are: to produce much less energy; to capture the resultant carbon dioxide before it enters the atmosphere; or to produce energy without burning carbon. Once this physics equation is embedded into the real human world of socio-economic and political systems, of course, things get more complicated. Anyone who argues that decarbonizing the world and fixing global warming is easy, is either a fool or a charlatan. This is the most complex problem that human society has faced. It's hard. In addition, we have also made it harder for ourselves – by which, I mean, vested interests in the rich world have made it much harder than it might have been for the rest of the world, particularly for the poorest in the global south, who are also the most vulnerable to a hotter world. We created this problem because we are humans with all the capabilities, flaws and marvels that entails; we will solve it only as humans.

There are lots of encouraging signs that the world is starting to act. To begin with, there is now near-universal acceptance of the human-made global warming crisis. In 2015, the same year the world reached 1°C of global warming above pre-industrial temperatures, governments meeting in Paris pledged to keep the temperature rise to well below 2°C and 'pursue efforts' to limit the temperature increase to 1.5°C by 2100. The Glasgow climate meeting in 2021 began a ratcheting up of national emissions-reductions pledges, and we've also made some other key steps towards meeting the Paris Agreement, most impressively with the phenomenal rise in renewable electricity production. It is now cheaper to install a brand new solar or wind power plant than to continue producing electricity at an existing coal plant. In the UK, renewable power production has already regularly exceeded that from fossil fuels. The plunge in cost of renewables has coincided with an accelerated improvement in their capabilities. We have better, more efficient solar panels, wind turbines, batteries and electric vehicles, and are far more savvy about integrating electricity generated in this way into grid systems. All this will only improve.

However, exciting as this progress is, it represents a mere fraction of what is needed even to stabilize emissions, let alone reduce them. To keep below 1.5°C of heating, we'd have to halve global emissions by 2025, and reach net zero by 2050. Instead, our greenhouse gas emissions are still growing (the annual increase continued even despite the major industrial shutdowns caused by the Covid pandemic), temperatures are rising, ice-melt is accelerating, and climate change is, as scientists predicted, getting worse. Carbon dioxide levels today are more than 50 per cent higher than the pre-industrial average.

Many scientists think it highly unlikely that we will stay below a heating of 2°C by the end of the century, let alone the 'safe' target of 1.5°C. Most countries are not making anywhere near enough progress to meet their pledged emissions reductions – and even if they fulfilled them to the letter, the national targets are so inadequate that it would be far short of what's needed to keep us below 2°C. Many countries significantly under-report their greenhouse gas emissions, so their climate pledges are anyway based on flawed data. China and India, the world's first and fourth biggest emitters, will have higher emissions in 2030 than 2020. In 2021, the Finnish town of Salla, located inside the Arctic Circle, launched its bid to host the 2032 Summer Olympics. The first ice-free summer for the Arctic Ocean is expected in 2035.

Climate models predict that we're on track for a heating of somewhere between 3°C and 4°C for 2100 – and bear in mind that these are global average temperatures. Subtract the seas from those calculations,

The global heating generation: How hot will it get in your lifetime?

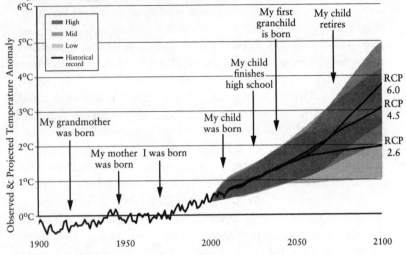

and what you find is that, at the poles and over the lands where people live, the increase may be double that, meaning people could be experiencing an increase of as much as 10°C by 2100. If that seems a long time away, consider how many people you know will be around then. My children will be in their eighties, for instance, perhaps with middle-aged children and grandchildren of their own. We are making their world. It will be very different.

Let's consider an entirely feasible planetary heating figure of 4°C by the end of the century, which is more likely than most people realize, so bear with me while I explain why. Climate modellers predict temperature rises based on various future emissions scenarios. The Intergovernmental Panel on Climate Change (IPCC) mapped out four different economic pathways (their term is 'Representative Concentration Pathway' or RCP) that we might take globally over the century: the 'RCP 8.5', in which we carry on business as usual with little attempt to decarbonize our economies; the moderate 'RCP 6.0', which sees emissions peaking by 2060 and then rapidly dropping; the intermediate 'RCP 4.5', which is more ambitious and sees emissions peaking by 2040 and dropping; and the very stringent 'RCP 2.6' pathway. Given current policies implemented since the COP26

How hot could it get?
UK Met Office estimates for global average temperature rise modelled according to the different emissions-reductions pathways the world's economies follow. The temperature could lie anywhere within the shaded range.

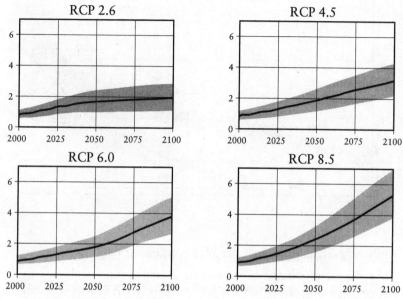

climate conference in 2021, we are charting a pathway somewhere between RCP 4.5 and RCP 6.0, with the former now slightly more likely. Projections show that 4°C by 2100 is somewhere between definitely possible and reasonably likely.[5] Indeed, we may well reach 4°C earlier than 2100, perhaps by 2075, even while sticking to the moderate pathway.

I'm using the UK Met Office plotted projections of change in global mean annual surface temperature (compared to pre-industrial temperatures) for the different emissions scenarios, since these factor in real-world systems, where things get messier. For instance, as soil heats up, biomatter decays faster, releasing more carbon dioxide into the atmosphere faster. The plumes represent the best estimate of the combined uncertainties in the modelling system, including things like cloud and water vapour feedbacks, which aren't included in the main IPCC projections. The plumes might get smaller as new understanding is brought in to the models – but on the other hand, there are

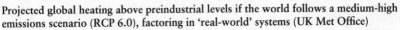

Projected global heating above preindustrial levels if the world follows a medium-high emissions scenario (RCP 6.0), factoring in 'real-world' systems (UK Met Office)

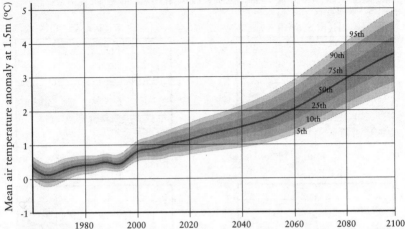

still things that are not fully included in a detailed way, like the effects of permafrost thaw and fires.

The rise in global temperature by a few degrees each decade will barely register for many people; the urgent problem is the extreme events the extra heat will trigger – heatwaves, flash floods, violent hurricanes and devastating fires. These are what overturn people's lives.

Alarmingly, there are signs that we may be tracking above the moderate pathway. Research published in 2021 found the melting of ice across the planet is accelerating at a record rate, with the rate of loss now in line with the worst-case scenarios of the IPCC.[6] About half of all the ice lost was from land, which contributes directly to global sea-level rises, with melting of the Greenland and Antarctic ice sheets speeding up the fastest, putting us on track for a sea-level rise of 1 metre by the end of the century. At the end of 2021, researchers reported huge and growing cracks in the Thwaites Glacier, a mass of ice the size of Britain or Florida that stretches across western Antarctica.[7] The floating shelf of the glacier could snap off into the ocean in as little as five years, they warned, triggering a chain reaction of collapse. A complete loss of the Thwaites could raise ocean levels by an additional 65 centimetres – or by several metres, if the collapse brings down surrounding glaciers with it. At least 28 trillion tonnes of land ice have been lost in the past twenty-five years – enough to put an ice sheet 100 metres thick across the UK, researchers calculated. And a loss of bright ice often exposes darker rock or ocean, speeding up further heating as the sun's heat is absorbed rather than reflected. In 2021, Arctic researchers announced that a significant part of the Greenland ice sheet is also on the brink of a tipping point, after which accelerated melting would become inevitable even if global heating was halted.[8] All things considered, a global temperature rise of 4°C by 2100 seems likely enough to plan for.

An average heating of the entire globe by 4°C would render the planet unrecognizable from anything that humans have ever experienced.

The world has reached a 4°C-hotter temperature before – well before humans appeared – some 15 million years ago, during the Miocene Era, when intense volcanic eruptions in western North America emitted vast quantities of carbon dioxide. Sea levels rose some 40 metres higher than today: the Amazon River ran backwards, California's Central Valley was open ocean, a seaway stretched from western Europe to Kazakhstan, spilling into the Indian Ocean, and lush forests grew in Antarctica and the Arctic. Atmospheric carbon dioxide rose up to about 500 parts per million, a level that today represents something close to the most ambitious and optimistic scenario possible for limiting our future carbon emissions. However, that global heating took place over many thousands of years, giving animals and plants time to adapt to new conditions and, crucially, the world's ecosystems had not been degraded by humans.

Things look bleak for our 2100 world, with plenty of extinctions as species struggle to migrate and adapt. In the oceans, we're looking at vast dead zones as pollutants combine with warmer waters to produce an explosion in algae that starve marine life of oxygen. On top of this, ocean acidity from dissolved carbon dioxide will cause a mass die-off of shellfish, plankton and coral – reefs will be lost well before 2100, at somewhere between 2°C and 4°C. Without coral reefs, which act as fish nurseries, fish populations will also plummet globally.

Sea levels will be perhaps two metres higher by 2100. We will by then be well on our way to an ice-free world, having passed the tipping points for the Greenland and west Antarctic ice sheets,[9] committing us to at least 10 metres of sea-level rise in the centuries beyond. By 2100, we will also have lost most other glaciers, including those that feed many of Asia's important rivers.

A wide equatorial belt of high humidity will cause intolerable heat stress across most of tropical Asia, Africa, Australia and the Americas, rendering vast areas uninhabitable for much of the year. Tropical forests of heat-tolerant species may well thrive in this wet zone, with the high carbon dioxide concentrations, especially with the disappearance of human infrastructure and agriculture, although the conditions will probably favour vine-like lianas over slower-growing trees.[10] To the south and north of this humid zone, bands of expanding desert will also rule out agriculture and human habitation. Some models predict that desert conditions will stretch from the Sahara right up through south and central Europe, drying out rivers such as the Danube and the Rhine.

In South America, models predict a weakening of the easterly trade winds over the Atlantic, drying the Amazon, increasing fires and turning the rainforest into grassland. The tipping point for the Amazon could well be triggered by deforestation; while the intact forest could cope with some drought because it generates and maintains its own moist ecosystem, areas that have been opened up through degradation allow moisture to escape, pushing it to a savannah state. By 2050, tropical rainforests, including the Amazon, may well be pumping out more carbon dioxide than they absorb.

This will undoubtedly be a more hostile, dangerous world. Heat will make large areas of the globe uninhabitable, and we will struggle to feed ourselves. Many of the places where people grow food will no longer be suitable because of heat stress or drought; despite stronger precipitation, the hotter soils will lead to faster evaporation and most populations will struggle to secure enough fresh water. Global food prices will soar, forcing tens of millions of hungry people onto the streets, into cities and across borders. Higher sea levels will make today's low-lying islands and many coastal regions, where nearly half the global population live, uninhabitable, thus generating an estimated 2 billion refugees by 2100, according to some forecasts.[11]

A 4°C world is a terrible prospect, uninhabitable for billions of people. I fervently hope we won't get there, but look where we are now at just 1.2°C above pre-industrial temperatures. Globally, it's already hotter than it's been in more than 100,000 years.[12] The only reason we don't already see a vastly different world, with forests flourishing in Antarctica, is that such changes take time. The world's systems are responding but aren't yet in equilibrium with the very recent changes to the atmosphere – that will take centuries.

We are leaving the sanctuary of an unusually stable climatic era in Earth's history – one which enabled crops to be grown and the flourishing of human civilizations.[13] We can already see its beginnings as we are thrown from one extreme weather disaster to another. Alarmingly, the global water cycle – the pattern of how water evaporates and is precipitated back down around the world – is now speeding up at twice the rate predicted by climate models and is likely to intensify by up to 24 per cent by the end of the century.[14] This would bring more intense and frequent hurricanes, dumping significantly greater volumes of rain. And it could well lead to a shift in key weather systems, such as the Intertropical Convergence Zone (ITCZ), a band of precipitation that hovers around the equator, where the trade winds meet in the tropics. The ITCZ feeds monsoons, and throughout history its position on the planet has meant life or, as befell the Classic Mayan civilization, perhaps death. Climate models disagree on how the ITCZ will react in our heating world, but we are heading towards a world of increasingly frequent and intense droughts and their opposite: deadly storms and floods.

Even reaching 1.5°C, expected by the early 2030s, is no picnic. At this temperature, around 15 per cent of the population would be exposed to deadly heatwaves at least every five years – that's 1.3 billion people, rising to 3.3 billion people at 2°C. By 2°C, bad harvests are twice as likely, and fish catches will have declined by twice as much as today. Sea levels are already rising faster than even the most pessimistic predictions.[15]

Our future world is one depleted of the biodiversity we're dependent on, and where we're facing a cocktail of negative shocks all the time, from fires to droughts. Within decades, we risk a turbulent, conflict-ridden world with great loss of life and perhaps the end of our civilizations.

2

The Four Horsemen of the Anthropocene

Climate change is a threat multiplier, which exacerbates the other social, environmental and economic problems that populations face. Fire, heat, drought and flood will transform our world this century. These four horsemen of the Anthropocene will make much of our world unliveable for people. Here's how.

FIRE

The first dawn of 2020 never arrived for my Auntie Helen on the south coast of New South Wales, Australia. The New Year's Day sky, black with smoke and ash, was lit with the same eerie orange glow through night and day. Helen spent the day packed onto a beach with hundreds of other evacuated householders, dogs, cats, horses and chickens, hemmed in by a wall of fire. The roads had become impassable and with flames audibly thundering around, the community had sought safety near water. But the flames kept coming. Eventually, they were rescued from the encroaching tide by boats.

A few days later thousands of dead birds washed up on the same beaches. Crimson rosellas, honeyeaters, rainbow lorikeets, robins, king parrots, whipbirds and yellow-tailed black-cockatoos, which had also sought refuge in the ocean, had succumbed to smoke inhalation and exhaustion.

While Auntie Helen had to be evacuated twice from her home in the space of a few weeks, Auntie Margi, her older sister, living in countryside in the north of the state, was evacuated just once, and with enough notice to grab precious belongings, photos and documents. Once the

This map shows regions where multiple severe impacts may occur at similar times at 4°C of global warming above pre-industrial levels. These impacts include extreme heat stress risk, river flooding, drought and wildfire risk, overlaid with an indicator of present-day food insecurity

Southeast Asia

Australasia

Southern Asia

Southern Africa

The Sahel

Sub-Tropical North America

Sub-Tropical South America

Number of High Severity Impacts
0
1
2
3
4

fires had passed through, Margi joined firefighting efforts near her home, wearing protective equipment and lugging water carriers on her 75-year-old back, up and down steep forest tracks to put out smouldering earth among the blackened trees.

'It was heavy, hard work, and it was scary,' she says. But she recognizes that this is the new normal. This is what it is to live with fire: the uncertainty, the ever-present rumbling stress, the bag packed, the reliance on community, the acceptance of bad air days, the declining value of your property and steep hike in insurance when that's even possible. After each devastating fire comes an erosion of the community. Some places are simply not viable, the fire risk too great. Small communities are erased from the map, plans for expansion denied. The places where it is safe for people to live are shrinking in Australia.

If it weren't for the global coronavirus pandemic, 2020 would have been the year we woke up – shockingly – to the Pyrocene, a planetary age of fire.[1] Millions of Australians spent the beginning of the year under a filthy pall of smoke, or confronting bushfires of a ferocity and scale never previously imagined. With hundreds of fire fronts burning, and record heat bearing down, more than 100,000 people were advised to leave high-risk zones in the nation's biggest ever evacuation.

Black Summer, as Australia's worst bushfire season has now been termed, was a direct consequence of climate change – 2019 was both Australia's hottest and its driest year – and the impacts will be felt for decades to come, even as such events become 'normal'. People living in the cities, far outside of forests, were overwhelmed by smoke, and suffered months of hazardous pollution. More than 80 per cent of the population was affected, thirty-four people lost their lives, and 6,000 buildings were destroyed. The true toll will be far higher, with estimates that smoke pollution led to 400 premature deaths and probable impacts on unborn and newborn babies too. Smoke killed ten times more people than flames. Australia is a country of migrants and its population is growing, but will people still choose to live in a country where a quarter to half of the year is spent battling intolerable heat and smoke?

The effect of the fires on native wildlife was horrific, and heartbreaking pictures bore global witness to kangaroos and birds attempting to escape, while tree-bound koalas shrieked as they succumbed to the flames. Nearly 3 billion wild animals were wiped out, making it one of the worst ecological disasters in modern history. The enormity of the devastation is so great that Australian scientists describe it as an omnicide, the killing of everything. Even forests populated by trees that flourish on cycles of burn and recovery are becoming less resilient in the face of bush fires that are growing in frequency, spread and intensity.

The extreme fires of Australia's Black Summer reflect a global trend across forests from California to British Columbia, Europe to Asia, the Amazon and Indonesia. Forests are naturally humid, but climate change is creating hot and dry conditions, more igniting lightning strikes, reduced winter snowfall and rain, and boosting invasive pests that turn vibrant trees into tinder. California experienced its worst ever wildfires in 2020, when more than 6,500 square miles burned, 100,000 people were evacuated, and around thirty people died.[2] In areas of high winds, some public utility providers had shut off power to prevent downed or damaged power lines from sparking, leaving families struggling with terrifying conditions in darkness. One-tenth of the world's giant sequoias are thought to have been destroyed by fire in 2020. Fire-weather days – days with high temperatures, low humidity and high wind speed – are projected to double in parts of the state by 2100, and increase by 40 per cent by 2065.

In 2019, fires in the drought-stricken Amazon produced so much smoke that the skies were darkened in São Paulo, thousands of kilometres away on the coast. In Europe, wildfires prompted evacuations in several countries, with record blazes in southern regions, such as Greece and Portugal. Nowhere is safe from the threat of fire – even wetlands have been burning. Neither are the world's coldest places immune. Arctic forests are burning, with mega-blazes devouring Siberia, Greenland and Alaska. Even in January, peat fires were burning in the Siberian cryosphere, despite temperatures below −50°C. These so-called zombie fires smoulder year round in the peat below ground, in and around the Arctic Circle, only to burst into huge blazes that rage across the boreal forests of Siberia, Greenland, Alaska and Canada. In 2019, colossal fires destroyed more than four million hectares of Siberian taiga forest, blazing for more than three months, and producing a cloud of soot and ash as large as the countries that make up the entire European Union. Models predict that fires in the boreal forests and Arctic tundra will increase by up to four times by 2100.[3]

Over the coming decades, entire national parks will burn in the US, with fire risk worsening across the west coast, but also increasing in wetlands such as the Great Lakes and Everglades, by as much as 500 per cent. Globally, models of fires show 'sharp increases projected for the European Mediterranean Basin and Levant, subtropical Southern Hemisphere (Atlantic coast of Brazil, southern Africa and central east coast of Australia) and southwestern USA and Mexico'.[4] The most massive fires are capable of generating vast energetic plumes – pyrocumulonimbus clouds – capable of injecting smoke up into the stratosphere where it can circulate around the globe for months, like the emissions of a volcanic eruption.

In addition to the devastation, fires also raise global temperatures, both by destroying vegetation (which draws down carbon dioxide) and by directly adding carbon dioxide from soils and combustion. The Black Summer fires produced up to 1.2 billion tonnes of carbon dioxide, equivalent to the annual emissions of all the world's commercial airliners. Smouldering fires are an even greater threat to the global climate,[5] because they burn for much longer, so they can transfer heat far deeper into the soil and permafrost, releasing twice as much carbon as normal fires.

Fires are increasing around the world. They are unpleasant, unhealthy, dangerous, expensive ... and people will choose – or be forced – to move away from them. Consider the pollution. Fire smoke, ash and unburned particles are hazardous to health, especially for people with conditions such as asthma. Later in 2020, during the sweltering heat of summer, friends of mine with young children living in Oregon, US, were unable to open their windows because of the thick smoke from forest fires there, and unable to stay with friends or family elsewhere because of Covid restrictions. In the end, the encroachment of fires forced them to flee, and the family had to live for several days in their car. Others who were evacuated, or who saw their homes or businesses destroyed by fire, had it worse. Some of those will rebuild and return – this time. Maybe next time too, but the time after that? For many, the decision will be made for them: insurance companies refusing to insure such risky properties;[6] government mandates banning rebuilding or habitation on hazardous sites. In January 2022, Adam McCay, director of the film *Don't Look Up*, an apocalyptic allegory for climate change, tweeted: 'Just had my home insurance cancelled because Southern California is at too high risk now for fire and floods.'[7] I personally know several Californian residents in similar difficulty – in some districts, the insurance commissioner has issued a moratorium on cancellations, but this is unsustainable in the long term.[8]

Friends talk of moving away from forested areas, country people move to the city where they have better access to fire services . . . More than a year after Black Summer, families in Australia were still homeless, surviving in temporary shelters in one of the world's wealthiest nations.

There is a lot that can be done to minimize fire risk through improved management. Ultimately, though, as the world gets hotter and drier, with more lightning, the risk of wildfires goes up. People will need to move away from fire-risk areas.

HEAT

Fire gets the headlines for its sheer power and violence, but its unglamorous cousin Heat is more deadly. The world now experiences twice as many days over 50°C than thirty years ago.

For most of history, the majority of people have lived within a surprisingly narrow range of temperatures in places where the climate supported abundant food production. As the world heats, this climate band is shifting away from the tropics, exposing billions of people to dangerous temperatures. Most of the global warming to date has been absorbed by the oceans: a mindboggling amount equivalent to around 20 zettajoules (20×10^{21} joules) of extra energy in 2020 alone,[9] or around ten Hiroshima atomic bombs a second. This is problematic for ocean life, because it slows vital mixing of the ocean layers that circulate nutrients and oxygen, and it is terrible for us, because it disrupts weather patterns and increases the likelihood of extreme, deadly events.

The oceans have done much to camouflage global warming on land, but it is nevertheless occurring, and fast. One in three of us could experience annual average temperatures of more than 29°C by

2070 – a climate currently found in only the hottest desert settlements.[10] We don't need to wait decades though; temperatures are already exceeding what modellers had expected for our current global increase of 1.2°C. In the summer of 2021, Death Valley recorded a crippling 55.6°C; Las Vegas was 47.2°C; even in Canada, temperatures reached 49.6°C, way above the decadal average. Freakish heat is also affecting the poles: in March 2022, Antarctica recorded temperatures more than 40°C hotter than seasonal norms at the same time as stations in the Arctic reached temperatures 30°C above the norm.

Heat becomes especially hazardous when combined with humidity, and we are already experiencing deadly levels that were not expected until 2050. For each degree Celsius the planet warms, the atmosphere holds about 6 per cent more water vapour. This becomes lethal because we can only cope with hot temperatures by producing sweat that cools us as it evaporates – high humidity means our sweat can't evaporate, so we overheat.

To measure the effects of heat plus humidity, scientists use what they call 'wet bulb' temperature calculations – meaning the lowest temperature to which air can be cooled via evaporation. At its most basic, this involves wrapping a wet cloth around the bulb of a thermometer and taking the air's measurement. Wet-bulb temperatures above 35°C, known as the 'threshold of survivability', will cause even fit people to overheat and die within six hours. Although that temperature might seem low, it equates to almost 45°C at 50 per cent humidity, and would feel like 71°C.[11] In the heatwave that ravaged Europe in 2003, for instance, the wet-bulb temperature hit 28°C, and more than 70,000 people died. In 2020 scientists discovered a handful of places – shorelines along the Persian Gulf, river valleys in India and Pakistan – had crossed the 35°C wet-bulb threshold for the first time in human history. Thankfully, they only did so for an hour or two at a time, but these events will become increasingly common.

By 2070, the earth's tropical belt will regularly experience temperatures as hot as the Sahara. Some 3.5 billion people live in this planetary girdle, which encompasses large parts of the Americas, Africa and Asia. By the end of the century, that tropical belt will be thousands of kilometres broader. By 2100, temperatures could rise to the point that just going outside for a few hours in some places, including parts of

Annual increases in extreme heat exposure for people living in urban environments 1983–2016

Person-days yr⁻¹
○ 10^{3-4}
● 10^{4-5}
● 10^{5-6}
● 10^{6-7}
● 10^{7-8}

India and eastern China, 'will result in death even for the fittest of humans under shaded, well-ventilated conditions,' according to one study.[12] Extreme hotspots include mid-latitude North America, the Mediterranean, the Sahel in Africa, and the rapidly desertifying deep Amazon in South America. Those living in these places will need to migrate to survive, and they will begin relocating well before 2070. Extreme heatwaves will occur multiple times each decade for billions of people by the 2070s.

Exposure to deadly urban heat has tripled since the 1980s, with a fifth of the world's population already affected.[13] In the next three decades, we'll see the band of subtropical climate expanding into higher latitudes by about 1,000 kilometres. London will start to feel like Barcelona, Moscow like Sofia and Tokyo like Changsha. To a Londoner, this sounds quite pleasant, but globally, it means that over 40 per cent of the world's forty-four largest megacities will see dangerously high heat conditions each year at just 1.5°C of warming, whereas at 2°C, 1 billion people would be exposed to extreme heat, rising to 3.5 billion people at 4°C, according to a 2021 UK Met Office analysis – and that doesn't even account for rising population levels.

At 1.5°C, EU member states alone will be experiencing 30,000 heat-related deaths per year by the 2030s. In 2010, when Russia experienced a mega-heatwave lasting nearly two months, bodies piled up in the city's morgues, devastating fires blazed out of control, and tanker trucks sprayed water to try to stop the asphalt on the roads melting. For more equatorial parts of the world, which are already experiencing deadly heat, such as the Indian subcontinent and parts of Africa, additional heatwaves will be catastrophic. Without wide access to air-conditioning, tens of thousands will die in an average summer, those labouring outside in fields or on roads and building sites being particularly affected. Countries in Africa could be spending tens of billions on air-conditioning to try to survive heatwaves in the +2°C world. The International Energy Agency (IEA) estimates that by 2050, air-conditioning demand will require additional electricity supply equal to the combined generating capacity of the United States, the EU and Japan. By 2100, climate change will be as lethal, per capita, as all infectious diseases combined are today, a 2020 study found.[14] The study's lead author states that a lot of older people already die due to indirect heat effects, describing it as 'eerily similar to Covid – vulnerable people are those who have pre-existing or underlying conditions. If you have a heart problem and are hammered for days by the heat, you are going to be pushed towards collapse.'

Quite apart from its impact on human and animal health, and agriculture, extreme heat also causes issues for infrastructure – buckling roads, railway lines and bridges, for instance. Globally, for every 0.1°C rise in temperature, the number of sinkholes increases by 1–3 per cent.[15] Heat even causes flights to be grounded, as planes struggle to take off in temperatures above 43°C.[16] There will be manifold problems societies face as we move suddenly into a very different world.

The date when scientists expect extreme heat to make parts of the world uninhabitable is also shifting closer because newer models now realize it will happen at lower global temperatures. Researchers initially put this threshold for mass migrations within a 5°C scenario; it's now between 3°C and 4°C, although even at a global temperature rise of below 2°C, at least a billion people would have to move.[17]

At 4°C, models show that exposure to extreme heat globally increases more than thirty-fold, compared to today, while in Africa it increases by at least a hundred times.[18] And it is not only the tropics; mid-latitude and even sub-polar regions will experience months of heatwave temperatures every year. Most regions within 30 degrees latitude of the equator may experience up to 250 days of extreme heat, resulting in a 'radical transformation' of the tropics and subtropics, which would spend much of the year in heatwave conditions, according to a 2018 study.[19] Half the Earth's land area and nearly three-quarters of the entire global population will be exposed to deadly heat for more than twenty days per year in a +4°C world. In the US, for instance, southern states will experience peak temperatures exceeding current Death Valley conditions every year, and more than eight weeks a year with temperatures above 37°C. Even the interior of Alaska will see annual temperatures in excess of 35°C. Consider what this means for major cities. New York City will suffer deadly heat levels for twenty to fifty days every year. Jakarta, though, will experience deadly heat *every single day*.

These temperatures will inevitably lead to an increase in deaths across the world – a rise in excess deaths related to heatwaves of 500 per cent projected for the US, and as much as 2,000 per cent in Colombia.[20] The most intense hazard from extreme future heatwaves is concentrated around densely populated regions of the Ganges and Indus river basins, a South Asian region inhabited by about one-fifth of the global human population. There, researchers say, wet-bulb temperature 'is likely to exceed the survivability threshold' in parts of northeastern India and Bangladesh, and to approach the critical threshold over most of the rest of South Asia. Close to a billion people will face a choice between risking death in each progressively hotter summer or moving. Meanwhile, researchers looking at the risk in China warn that heat and drought 'may limit habitability in the most populous region [the North China Plain] of the most populous country on Earth'.[21] Deadly heatwaves and dangerous wet-bulb temperatures are expected across the North China Plain and eastern coast of China, even in an emissions reduction scenario (RCP4.5), over the next three decades. This is a region of at least half a billion people and includes the major cities of Shanghai (population 34 million) and Hangzhou (22 million).

You may think we can overcome the difficulty of heat and humidity through air-conditioning and desalinated water. After all, impossible cities like Dubai and Doha in desert regions do just that – Qatar has even begun air-conditioning the outdoors, cooling sports stadiums, pavements and outside markets and dining areas, a trick that has helped it become by far the world's biggest per capita greenhouse gas emitter.[22] So yes, in some places some populations will be able to endure for longer by essentially living an indoor, cooled or nocturnal existence, and perhaps by wearing powered, air-conditioned bodysuits.[23]

However, even if you ignore the energy and water costs of extreme adaptation, such strategies will only ever work in a limited way for a small section of urban society. After all, those populations would be heavily dependent on resources produced outside, such as food. Consider that even for wealthy Gulf states, there are continual fears over food security, as there are no permanent rivers or lakes in the region, so vulnerability to shocks, such as the Covid-19 crisis, is acutely felt. Countries like the UAE import 90 per cent of their food. Right now, half of the world's food is produced on smallholder farms using physical labour, yet as the world warms there will be more and more days when it will be physically impossible to work outside, reducing productivity and food security. Rice farmers in Vietnam are already planting at night-time with headlamps to avoid dangerous heat,[24] and Qatar had to impose a ban on outdoor labour between 10 a.m. and 3.30 p.m. as early as May. As *The Lancet*'s climate commission has shown, in 2018 already more than 150 billion work hours were lost due to extreme temperature and humidity.[25]

This number could double or even quadruple depending on how many people keep working in rural agriculture until it becomes economically unfeasible and impractical to continue. The International Labour Organization calculates that under the most optimistic global warming scenario, of only a 1.5°C temperature rise over this century, the increase in heat stress will lead to global productivity losses equivalent to 80 million full-time jobs in the year 2030. This is equivalent to global economic losses of US$2.4 trillion. This is a conservative estimate, based on outdoor jobs, such as agriculture and construction, being conducted in the shade – which self-evidently they are often not. Already the richer, more climate-controlled parts of the world outsource much of their hot, dirty work to the poor, hotter parts. People suffer heat stress, dehydration and exhaustion in overcrowded, sweltering factories and sweatshops to produce the ultra-violet protecting shirts and air-conditioners, and to construct the marble-clad hotel lobbies that keep those in the rich world cool.

These changes will exacerbate existing social inequities. Even in wealthy nations, it is often migrants from poor nations that work in unbearable heatwaves to harvest crops in the fields. Poor, densely populated neighbourhoods are generally hotter than rich, tree-lined ones. Women and girls are more likely to die in heatwaves than men. Indeed, multiple studies show that women are more vulnerable to disasters, including to the impacts of climate change: they are more likely to be displaced during extreme events, and to lose their jobs or have their income slashed; and girls are more likely to lose their education. More than 60 per cent of women work in agriculture, where they experience a greater workload than men and are less able to adapt because they have less access to information than men.

Inequality kills. Studies in the US show that within the first ten years of life, heat has a huge and deleterious impact on everything from health to education, but those living in the poorest neighbourhoods are far more affected.[26] Poorer neighbourhoods, predominantly occupied by Black and Latino people, are on average 2.8°C hotter than rich ones within the same US cities, and those households are half as likely to have aircon units, prolonging their exposure to heat. The racist housing policies that led to this disparity even have an impact before birth, because heat is a risk in pregnancy,[27] and its dangerous effects can be seen particularly in poorer, Black communities. It will only get worse over the coming decades, since the southern US will experience some of the worst impacts of heat and climate change. Heat, like so many climate change impacts, is mediated through inequality.

DROUGHT

As the planet heats, despite the increase in humidity, more of the rain will fall over the oceans and less on land. We may well have already moved into an era of mega-droughts; by the time global temperatures rise by 4°C, we will be in a Dustbowl world.

The hundreds of millions of people who rely on mountain glaciers for their water, particularly in South Asia and South America, risk the loss of entire breadbaskets once these water stores are gone. At least 129 million people in South Asia are substantially dependent on upstream meltwater for their livelihood, in addition to 221 million people in Pakistan, Afghanistan, Tajikistan, Turkmenistan, Uzbekistan and Kyrgyzstan. Models show that parts of this region will hit 'peak water' this decade, with a subsequent sharp decline through the rest of the century as the glaciers wither and die.

By 2050, models show a reduction in annual rainfall across the Mediterranean, Australia and southern Africa. The most severe reduction is projected for the top half of South America, encompassing much of Brazil and the surrounding countries, and covering virtually the entire Amazon rainforest, where researchers predict the most intense drying of anywhere in the world.

Of all the global changes that I've witnessed, drought has been the one that's affected the most people. On my travels to remote parts of the world, I've watched rural villages depopulate and die as drought wipes out agricultural livelihoods and precious food; in the sprawling shanty towns of the world's expanding cities – from Mumbai to Nairobi to Lima – I've seen the consequences of these deceased villages. In Uttarakhand, for instance, a mountainous state in northern India, more than 4 million people have migrated (40 per cent of the population), leaving nearly 800 uninhabited 'ghost villages', as temperature rise and drought have made agriculture at altitude almost impossible. Across South America, from Peru to Bolivia to Colombia, rural villagers are experiencing the combination of long-term crippling drought and the disappearance of the Andean glaciers that used to provide meltwater for irrigation.

Even twenty years ago, the village of Overjeria, in the Bolivian highlands, was a busy farming community whose maize, quinoa, potatoes, avocados and fruit were sold in the markets of the capital La Paz. By 2010, climate change had ravaged the landscape. Persistent droughts killed their crops, their livestock and, finally, the village. When I visited, there were just nine old people left, surviving in adobe huts. Luciano Mendez, a 75-year-old, ancient-looking farmer, who sucked on a cheekful of coca leaves to numb his hunger, told me the last of his eight children had left with his family three years earlier, after multiple failed harvests. 'We only get twenty minutes of rain every few days, during the rainy season,' he told me. 'The cows die first, then the donkeys. Goats are the hardiest.'

The younger men go first – some mere children – then whole families, leaving their Andean villages for towns and cities. They're easy to spot, all the way up to Colombia and into Central America, with their belongings wrapped in Quechua shawls over their shoulders, weary from weeks of sleeping rough. It is no surprise that this continent leads the developing world in rural to urban migration. Subsistence farmers, for whom one bad harvest can mean unbearable hunger, are experiencing reduced yields in most key crops from rice to wheat.

One *Lancet* study calculates that a global temperature rise of 2°C would lead to a reduction (compared to today) of global food availability in 2050 of 99 kilocalories per person per day – a serious issue for people on the edge of starvation. Food crops will also become less nutritious under heat-stressed conditions, with reductions in protein, zinc and iron of almost a fifth compared to currently. The human impacts could be dire. 'We estimate that elevated carbon dioxide could cause an additional 175 million people to be zinc deficient and an additional 122 million people to be protein deficient,' researchers warn.[28]

The problem is that our food comes directly or indirectly from plants, and plants need water to grow. As we heat the world, water evaporates faster from soil and leaves, and it doesn't rain as regularly or as much. In addition, heat-stressed plants (and animals) need more water. What this means is that as global temperatures rise, agriculture will become increasingly difficult, and in many places impossible, forcing farmers to relocate. Given humans cannot live without food, we're going to have to come up with solutions fast.

Heat itself is damaging to crops. Plant cells, tissues and enzymes get destroyed at about 39°C, often killing the whole plant. Each day spent above 30°C reduces maize yields by 1 per cent, and by closer to 2 per cent under drought conditions, so a three-week heatwave could slash yields by a quarter. Studies show that with a 4°C global temperature rise, where heatwaves will push regional temperatures to the high forties in mid-latitudes, and the high fifties in the sub-tropics, crop losses will be unsustainable.

The US stands to lose half its maize crop, for instance, including most of the current-day Corn Belt. When droughts are factored in, losses soar to 80 per cent or more. This won't be just a domestic catastrophe: the US, Brazil, Argentina and Ukraine account for nearly 90 per cent of global maize exports, and a 4°C rise in global temperature would slash harvests and eliminate any exports. Wheat, which globally provides a fifth of all our food calories, is similarly threatened – in recent decades drought impacts on global wheat production have doubled. By 2050, severe water scarcity is expected in an almost continuous belt from the Iberian Peninsula in the west to Anatolia and Pakistan in the east, across southern regions of Russia, the western United States and Mexico.

Before the end of the century, more than half the world's land surface will become classed as 'arid', models show. More than three-quarters of the affected areas are in the developing world, but hyper-arid areas appear even in Alaska, northwest Canada and Siberia. Even areas not classed as drylands will experience more frequent and severe extreme droughts, including the whole European continent, with the exception of Iceland. Globally, an additional 3 billion people will suffer water stress by 2100, with a third of the world's population no longer having access to sufficient fresh water – something that will also impact sanitation, increasing the risk of ingesting pathogens.[29]

Large portions of the globe will be unsuitable for agriculture, including livestock farming. When rural livelihoods become impossible, people have to move.

FLOODS

While a huge swathe of the globe, including the world's most populated regions, will become unliveable because of too little water, people living in other places – including many of the world's biggest cities – face the opposite problem: too much water.

There are three main ways a hotter world threatens us with too much water: hotter oceans mean more expansive oceans with higher sea levels; hotter land means melting ice bringing flash floods to rivers and deltas, and additional sea-level rise at coasts; and a hotter atmosphere means more violent storms and extreme precipitation. All of these threaten people, particularly those living in low-lying areas, near coasts and river systems – in other words, the vast majority of humans. Sea levels, for instance, are already rising faster than forecast, and could exceed 1 metre by the end of the century. This would be a catastrophe for our cities, most of which are coastal and home to hundreds of millions of people.

A primary risk of rising sea levels is groundwater becoming salinated. We are already seeing the impacts on agriculture in Bangladesh, for instance, where rice farmers have had to convert their fields to shrimp aquaculture or abandon their lands to work in the textile industry of Dhaka. And higher sea levels lead to greater storm damage and coastal erosion, forcing more communities to abandon uninsurable homes and rural livelihoods that had become impossible. All of the slum dwellers I spoke to in Dhaka had fled their villages for this reason.

The prospect is especially bleak for low-lying islands and atolls, including the nations of Maldives and Tuvalu, which will be made uninhabitable from as soon as 2050, through increasing erosion and seawater submersion that infiltrates groundwater and destroys soils, vegetation and infrastructure. Current international law measures a country's exclusive economic zone – including fishing, mining and tourism rights – from its coastline, so as that retreats or disappears, so too do its oceanic economic territories. This double whammy means their terrestrial and marine economies are at risk simultaneously.

Even if global temperatures rise no more than 1.5°C, hundreds of millions of people will be affected. There are a significant number of countries where at least 50 million people live on land exposed to sea-level rise – these include China, Indonesia, Japan, the Philippines and the United States. Once global temperatures rise by 2°C, flooding will affect at least 136 megacities, with damage costs totalling $1.4 trillion per year by the end of the century. The cost of holding back the rising ocean, through sea walls and other defences, is already prohibitive – consider that the US alone estimates spending more than $400 billion in coastal defence costs over the next twenty years, with longer-term costs for those living in smaller communities as high as $1 million per person.

On average, 1.7 million people will be displaced for each centimetre of global sea-level rise, so hundreds of millions of people would have to move by 2100. The whole of southern Vietnam is expected to be below sea level by 2050, as well as much of the central and northern parts of the country.[30] Coastal Florida is already seeing the signs of a climate-driven housing crisis, with first sales tumbling and then prices, as luxurious waterfront homes become too risky to purchase. Sales in vulnerable areas were down 20 per cent on safe areas in 2018, after Hurricane Sandy, which damaged some 650,000 homes and left 8.5 million people without power (some for months), made buyers more wary.

Europe, with its densely populated 100,000-kilometre-long coastline, will also be seriously affected. The number of people exposed to coastal flooding is projected to rise from 0.1 million a year today to 3.6 million by 2100, with the UK worst hit financially, followed by France and Italy. Bear in mind that the Dutch already spend €1.2–1.6

billion per year on their Delta Programme to protect the nation from flooding, and London and Venice have invested substantially in protective storm-surge gates. Protecting coastal cities could cost perhaps hundreds of billions of euros per year, and as rising waters increase the numbers of people living below sea level, a breach of those defences would then risk cataclysmic losses. In the longer term, over centuries, none of today's coastal cities will be viable in a 4°C world, so the question will be when and how to abandon them. As a team of scientists wrote in 2016, humanity 'will have a very limited time after midcentury to adapt to sea level rises unprecedented since the dawn of the Bronze Age'.[31]

Flooding will increasingly be a problem away from the coasts, too, because as the global temperature rises, air holds more water and becomes more energetic. This means extreme weather events will increase in frequency and severity, dumping catastrophic amounts of water, wiping out crops, houses and roads, and causing great loss of life. Extreme precipitation in the South and East Asian monsoon regions is the most sensitive to warming, scientists found, with roughly a 10 per cent increase in intensity for each degree of temperature rise. By 2050, one-fifth of the world's land area will see a significant increase in severe week-long flooding, with poor countries most seriously affected. Bangladesh faces an increasingly perilous future, overwhelmed by rising seas from the south and powerful river flows from the north. Today's once-in-a-century flood flows are projected to increase by 80 per cent for the Meghna, 63 per cent for the Brahmaputra and 54 per cent for the Ganges by the end of the century, and the three rivers are also more likely to synchronize their times of peak discharge.[32] In addition, storm surges from stronger cyclonic storms in the Bay of Bengal mean that less water will be able to flow out to sea. Bangladesh is one of the most densely populated nations in the world, so this means that tens of millions of vulnerable people face regular or even near-permanent inundation.

In March 2022, my bushfire-surviving Auntie Margi was again cut off from the world and rationing out her remaining tinned foods – this time she was victim to horrendous floods that hit the east coast of Australia after a 'rain bomb' weather event that lasted days. Margi's house in Lismore was damaged; many of her neighbours' houses were completely destroyed, and the town centre was unrecognizable. She sent me photos of the community centre floating down a fast-moving river. People had to be rescued by helicopter from rising water and landslides. Hundreds of thousands of people faced evacuation orders, and with so many homes destroyed, not everyone could return.

Flooding will become an increasing problem far outside monsoon regions as well, with heavier rainfall increasing river flows by as much as 50 per cent across the Northern Hemisphere. The many towns and villages that are already flood-prone, and properties built in flood zones, will be abandoned and uninsurable, as the at-risk zones grow. This applies to large parts of cities too – many of the people who died when Hurricane Ida hit New York City in 2021 were poorer residents living in basement apartments that flooded.

I live in a 125-year-old Victorian house on a hill in a suburb of London. My own home will be safe from flooding, according to the hydrological projections, but the many houses, schools, shops and transport links at the bottom of my hill, by the local river, will not. Analysis by the office of London's mayor reveals one-fifth of London's schools are susceptible to flooding in the next decades. Glad as I am that my own floors will remain dry, none of us can function marooned from society, and floods all around will be distressing and expensive. The roads around me – two of which are known to have been there since Roman times – will need to be raised; so too the century-old railway line. And that's just one tiny part of one borough of one city.

Today's tropical weather systems, such as hurricanes and cyclones, will become more frequent and severe, and increasingly also occur at higher latitudes. Once global temperatures rise by 4°C, models predict twice as many strong El Niño events, capable of displacing rainfall belts by as much as 1,000 kilometres and triggering weather chaos around the world. During the 1997–8 mega-El Niño, catastrophic floods occurred in South America in normally arid areas, marine life was devastated, and 23,000 lives were lost in weather-related disasters. Extreme La Niña events often follow extreme El Niños. When this occurred in 1998–9, the southwestern US experienced one of its worst-ever droughts, flash flooding and landslides killed some 50,000 people in Venezuela, storms and floods displaced 200 million in China, and half of Bangladesh was submerged by flooding. La Niña can also favour an intense North Atlantic hurricane season, which in 1998 saw the formation of Hurricane Mitch, one of the deadliest and strongest storms on historical record, claiming more than 11,000 lives as it struck Honduras and Nicaragua in Central America. Extreme events like these will become increasingly regular, particularly in the low latitudes in a wide belt around the equator, forcing millions from their homes in search of safety.

A liveable planet is not a lost cause. It is still within our agency to turn this around and we must try. Every degree of temperature rise we avoid, the safer we will be; every tenth of a degree matters.

That said, we're already at a mean global temperature rise of 1.2°C, so keeping below, say, 1.5°C means acting hard and fast. Even if we were to stop all our global warming emissions today, the inertia of the Earth's climate system means that global temperatures would continue to rise for a few years before falling. In our favour, though, this inertia – systemic time lag – means that we do still have time to rapidly reverse course. And global leaders are certainly talking seriously about climate action – the two biggest polluters, the US and China, have committed to reaching net zero emissions by 2050 and 2060 respectively, for instance. If all the world's net zero pledges were met, the global temperature rise could potentially come down as low as 2.1°C by the end of the century. This is a significant challenge, and there is little evidence to suggest we're preparing to meet it. To have any chance of limiting the temperature rise to 1.5°C, we'd have to massively increase our short-term ambitions. There are ways to do this – we must – and I will look at those later in the book.

Nevertheless, we need to be realistic, and the most likely scenario, at time of writing, is for global temperatures to rise by 3–4°C this century even while our mitigation efforts prevent still-greater planetary heating. As I've outlined, this makes large swathes of the world uninhabitable – places where most of the world's population currently lives. There's no magic date for when this occurs; the world is already a more dangerous place for millions of people because of the current level of emissions, and it will only get worse with every rise in temperature. Initially, this will continue to be a far bigger problem for the poorest, who are most dependent on their immediate environment, who live in the worst-affected regions of the global south, and who cannot insulate themselves from the changing conditions with aircon and flown-in meals. Before too long, though, it will be a problem for everyone.

Compounding these problems is the fact that the global population is still growing, particularly in some of the regions worst hit by climate change and poverty. Populations in Africa are set to quadruple by 2100, even as those elsewhere slow in growth. This means there will be a greater number of people in the areas affected by extreme heat, drought and catastrophic storms – and a greater number of people will need food, water, power, housing and resources, just as these become ever-harder to supply. Globally, population is projected to peak around 2064,[33] before declining to around today's level by the end of the century. This addition and then elimination of billions of people, during a time of enormous planetary disruption in our climate, ecosystems and water availability, adds considerable pressure to humanity's adaptive capabilities.

We face a very hostile world, characterized by a belt of uninhabitability swathed across most of today's most populated regions, including much of Asia, Africa, Latin America and Oceania. This is a completely new situation for our species, one in which our expanding population must deal with an ever-shrinking zone of habitability, within the restrictive cage of social and geopolitical boundaries.

Although the scale and extent of what we face this century is unique, we have over the past hundreds of thousands of years experienced other crises. We've survived them by migrating. Migration is not the problem; it is the solution – it always has been. As we will see, migration is the oldest survival trick.

3
Leaving Home

Migration is our way out of this crisis.

Migration made us. This might be hard to see in the context of today's geopolitical identities and constraints, where it can feel like an aberration, but, viewed historically, it is our national identities and borders that are the anomaly. Migrations, whether for exploration and adventure, from disaster to safety, for a new land of opportunity, for god and soul, for trade or art, under duress and by kidnap, have transformed our globe and globalized our species. Human migration fundamentally created the human system we are all a part of today.

For humans, migration is deeply interwoven with cooperation – it is only through our extensive collaborations that we are able to migrate, and it's our migrations that forged today's collaborative global society. Only by understanding the peculiarities of our species, and how we got to dominate the planet and its climate, will we be able to see our way forward – embracing our inherent strength of collaborative migration to thrive through the environmental crisis we face.

Migration is a survival strategy used widely in nature. Many species have evolved an innate migratory response to the seasonal or geographical variations in food and weather conditions. Animals as diverse as the bar-tailed godwit and the Atlantic salmon undertake long, arduous and dangerous journeys because their biology compels them to do so. I can certainly relate. In the bleak beginnings of 2021, after several months of pandemic travel restrictions, during which I barely left home, I felt myself overcome by a form of *Zugunruhe*, the 'migratory restlessness' that seizes birds when it is time to take flight. The signs are obvious: sleeplessness, disruption of normal activities, agitation ... A caged American robin will launch itself northwards again and again, hammering against glass walls even if it has no view of the outdoors. A western sandpiper's digestive organs atrophy to accommodate the demands of migration,[1] and they obsessively stock up on specific foods and even performance-enhancing drugs.[2]

We, too, are a restless species. Most of us choose – indeed, we pay – to spend time away from home. We do so, not to gather essential resources and food, but to enjoy the Elsewhere. Even if we don't live nomadic lifestyles any more, or migrate permanently, we retain this desire and curiosity to explore new places and to live even for a few days in a different location.[3] It is unusual – abnormal even – to remain in your home and leave only for essential trips. Indeed, it is associated with agoraphobia, a psychiatric disorder.

Studies across nature show that global dispersal is the most effective strategy for any species to prevent extinction. However, only a few species are able to adapt to multiple environments; most are exquisitely adapted to their particular niche. Among primates, only humans have spread across the globe without evolving into different species at different locations.

At some point in our hominin ancestry, we gained the ability to live anywhere. However, in gaining this superpower, we lost any innate adaptation to a particular environmental niche. For a creature as big and demanding of calories and resources as a human, this was an evolutionary risk. It was only possible because human brains are highly adaptive and we are hypersocial, able to cooperate with large numbers of unrelated people, supporting each other and sharing resources, ideas and knowledge. Humans learned how to change the environment to suit their needs, and they also learned how to adapt themselves with skills and behaviours to suit different environments. Migration allowed our species to survive environmental challenges, intertribal clashes, territorial disputes, food and resource scarcity, inbreeding and disease.

When climate allowed, hunter-gatherers would eventually people the world. But these were not solo migrations. Humans relied on group collaboration, building up webs of cooperative agents to spread the risk and the energy involved in moving far from their original evolutionary niche. This, after all, was the Pleistocene, and a very different world to the one we know today. Mile-thick ice sheets cloaked northern Europe and Asia, and one-third of North America, and vast numbers of large mammals roamed the ice-free landscapes, including dozens of fearsome carnivores, long since hunted to extinction. Navigating such a landscape required flexibility and a deep reliance on group support. We will need both this century.

THE MIGRATION OF STUFF

We don't just move ourselves, though. Look at where you live now: you can only survive in the land you're occupying – your new 'niche' – because of all the bits of other environments that have been brought to you. In my case, this includes all my food, water, every piece of infrastructure and object in my home. The only resources I am using that are derived from this environment, are the actual ground I'm sitting on and the air I breathe. I'm dependent for everything else on an intricate global network of thousands of people who are moving around the planet and moving the stuff of the planet around to me.

Human migration depends on a key secondary migration: the movement of stuff. For our early ancestors, this included water (carried in pouches), enabling them to work as endurance hunters over days, and the tools they needed to kill, cut and process prey, such as stone axes, wooden spears and kindling for fire-making. The ability to carry the resources they needed allowed them to bridge sparse, inhospitable areas and make far longer journeys. We are the only animal that does this and it liberated our ancestors, enabling them to develop their technologies over a lifetime and over generations.

The next step was to combine collaboration with resource movement: trade began within and between groups. Trade dramatically reduced the energy costs of resource acquisition especially on long journeys, and reduced the risks of starting life in a new location. By the time of our *Homo* ancestors, people had created social networks between groups that were strong enough to enable exchanges of resources far from their geographic origin and between people. It was this that gave them the ability to migrate further and longer, eventually leaving Africa to colonize Asia, Australia, Europe and the Americas.

It was perhaps this ability to exchange resources among supportive networks that gave our species of human the edge over other now-extinct ones. When modern humans displaced Neanderthals, population densities increased at least tenfold. The main way they increased the land's carrying capacity may well have been through wealth transfers made using a 'currency' of valuables, such as shell beads. Neanderthals also made a range of decorative items but it is not clear that they traded widely. Our ancestors, however, collected and traded raw materials over great distances, and from them made items of added value that they also traded. Trade – the organized migration of resources – allowed our ancestors to build greater social networks and increase their group size, cultural institutions and their resilience to harsh environments. Trade helped groups to specialize in specific cultural activities and technologies while still being able to meet all of their needs. It enabled our ancestors to occupy land across continents, whereas Neanderthals never ventured past Eurasia.

Today's hunter-gatherer tribes often separate into bands during the hunting season, and then come together at great festivals for a week or so, a few times a year. During these gatherings, meat, stories and other resources are exchanged, ideas, techniques and tools are pored over, decorative items are examined, and trading relationships are developed. In preparation for these festivals, modern-day hunter-gatherer societies, like the !Kung peoples of the western Kalahari, devote substantial time to making tradeable valuables, such as ostrich-shell jewellery, which they trade for migratory rights to another group's territory to hunt or gather food there. Trade helped ancient groups to migrate because it spreads environmental risks – if the water holes dried in one tribe's territory, causing a dearth of game, then it would be possible to acquire food in an exchange from another tribe further off.

Human migration for early hunter-gatherer ancestors was continual but incremental. The hostile environment of the Pleistocene era, where humans spent most of our evolved history, kept populations low and limited opportunities for exchanges between groups, and this is reflected in small genetic differences that subsequently arose between separated descendants of the same relatively small group of African explorers. Skin colour, controlled by several different genes, is a visible marker of ancestral migration, generally lightening (losing melanin) with latitude, as the sun's strength weakens. Melanin protects against ultraviolet rays, but limits the amount of essential vitamin D the skin can make in its reaction with sunlight. While chimpanzees have light skin, the default for hairless humans is protectively dark. The familiar pale-skinned European emerged incredibly recently; Europeans had dark skin and hair (with blue eyes) even 4,000 years ago.

Before the end of the last ice age all humans migrated, settling for no longer than a season, meeting up in regular gatherings of tribes, but soon dispersing into more sustainable population densities. The first Britons, for example, didn't settle. They came to hunt and then retreated back across a land bridge to the warmer lands further south on the continent.

HOW THE MIGRATION OF STUFF GAINED PRIMACY

Migration diversified our genes and our culture. Take northern Europe: there were three key mass migrations during the Stone and Bronze Ages that dramatically altered the genetic make-up of all Europeans, mainly because they took place at times of low population. The first of these, arriving from around 18,000 years ago, was hunter-gatherers migrating north from the Balkans, whose DNA still makes up about 30 per cent of European genes. The second was a northerly migration from Anatolia of farmers who arrived around 8,000 years ago,[4] and initially lived parallel lives with the indigenous hunter-gatherers. The third, which occurred as recently as 5,000 years ago, was the migration of nomadic pastoralists from the Eurasian steppes into settled farmlands.

It was the second migration that began fixing people to geography – to plots of land. The ancient transition from a lifestyle of hunting and gathering to permanent human settlements put extra pressure on local resources. Growers radically changed the planet's landscape, making it more productive for human survival. Once we invested in a field of sown crops, we needed to be around to reap the benefits, so agriculture fundamentally changed human migration patterns. We became tied to the land – even if we didn't own it, our identity was linked to it. However, agriculture at first only made us relatively sedentary – as we depleted landscapes, we moved on.

Today, climate change is uprooting us, but back then, it was key to a different change – you could say, it was rooting us. What I mean is, this transformation of our species into farmers was only possible because of global climate change. The earth's atmosphere during the last ice age contained so little carbon dioxide, around 180 parts per million, that photosynthesis was very inefficient, and consequently the planet's total vegetation was just half of what it is currently. Nomadic tribes living 20,000 years ago would not have been able to permanently settle, because the spindly wild grasslands that managed to grow at that time couldn't have supported permanent herds, let alone farmers. By 8,000 years ago, however, atmospheric carbon dioxide had risen to 250 parts per million, leading to a phenomenal rise in plant productivity, so that hunter-gatherers didn't need to travel so far for supplies, herds could settle for longer periods, and settlements had enough stability to invest in infrastructure, from irrigation channels to protective silos for grain.

However, farming, especially early on, was an insecure lifestyle and many starved or lived on the edge of survival – local wildlife would have been sparse as settled humans depleted it, and if harvests failed, migrating to new pastures was harder. A key tool in their survival kit – the ability to migrate – had been broken just as they most needed it. Evidence from an archaeological site in Anatolia dating to between 9,100 and 8,000 years ago, for example, shows that as the population rapidly increased, so did its rates of bone infection and tooth decay from impoverished starch-based, low-protein diets. Despite these health disadvantages, agriculture is the most efficient way of using land for food, and the sedentary lifestyle meant women could reduce spacing between births, as they weren't limited by the burden of carrying babies and toddlers. They thus had more children, who required more land. So agriculture expanded human populations and dispersal in a new and important way.

Demographers have modelled the evolution of sedentism and concluded that it only occurs where population reaches a level that impedes constant movement (and also the resource depletion rate is low).[5] In other words, sedentism and agriculture evolved out of our success as a migratory species. It was agriculture that enabled civilizations and allowed them to flourish. Social well-being took a hit, though, and many injustices persist today. There is evidence that egalitarian societies became far more socially unequal through settlement. The extraordinary settlement of Çatalhöyük in today's Turkey – already a city, 8,000 years ago, of hundreds of mud-brick homes – reveals evidence of a remarkably egalitarian society. By 6,500 years ago this had changed, with inequality among households and violent punishment for wayward members of society – as evidenced by skulls with deliberate attack marks that had healed.

Settlement had clear advantages at the societal level, allowing larger populations that could develop more complex cultures. But being fixed to one place makes you vulnerable if the conditions become unsafe. Climate change, for instance, has always prompted migration. Whenever the planet has cooled, people have migrated south towards the equator; whenever it has warmed, people have pushed north. The shocking and brutal ice age of 12,900 years ago, when parts of Europe dropped 12°C in fifty years, sent hunter-gatherers south to the Middle East. The global warming that began slowly some 10,000 years ago set them mobilizing north.

Everywhere agriculture was invented or imported, farmers colonized land formerly roamed by nomadic peoples. The first African farmers sowed what is now the East African Sahara some 8,000 years ago, but the climate changed 6,000 years ago, drying the region and expanding the desert, in a rare case ending agriculture there in favour of nomadic pastoralism and hunter-gathering. Around 4,500 years ago, the Bantu people of West Africa domesticated the yam and began large-scale migrations westwards and southwards, driving the formerly sizable populations of Khoisan Bushmen and Pygmies out of fertile lands and into small remaining populations in the savannahs and forests respectively. Similarly, in the Americas, the Toltecs and Aztecs were farming migrants who replaced indigenous communities.

Settled farming enabled populations to increase, as we've seen, but there was a limit. For thousands of years, farmers relied mainly on recycling biological matter to make up for the shortfall in the nitrogen and other essential nutrients available to crops in the soil. They left the stalks and silage in the fields to rot, added whatever other organic material they could, including animal and human manure, and practised crop rotation. But as populations grew, more crops had to be provided from the same area. Then, in 1909, the German chemist Fritz Haber invented a way of turning the nitrogen in air into a form that plants can absorb; his colleague Carl Bosch scaled it into an industrial process. The era of artificial fertilizers was born: the effect on population growth was immediate. Half of the protein in our bodies now comes from the Haber–Bosch process. Billions of people owe their daily bread, rice and potatoes to artificial fertilizers, which have utterly transformed the planet's nitrogen cycle and helped catapult our species into global dominance with a population of 8 billion.

The global migration of food produced with modern agriculture has allowed the vast majority of us to settle in the same small areas that have been occupied for hundreds of generations. These are areas with large concentrations of people, which produce almost no food but rather trade in migrated foods and other resources from outside.

Agriculture was the invention that turned a migratory human into a settled one. Although it was never that simple, of course. Humans didn't simply stop migrating after we stopped chasing our food around. For one thing, agriculture is a precarious business that puts the calories of entire populations at the mercy of environmental conditions. Around 3,200 years ago, for instance, an entire network of civilizations collapsed when climate chaos triggered a 300-year period of drought in the Near East.[6]

'There is famine in [our] house. If you do not quickly arrive here, we ourselves will die of hunger. You will not see a living soul from your land.' This letter was sent between associates at a commercial firm in Syria with outposts spread across the region, as cities from the Levant to the Euphrates fell. Across the Mediterranean and Mesopotamia, dynasties that had ruled for centuries were all collapsing. The walls of Ramses III's mortuary temple – he was the last great pharaoh of Egypt's New Kingdom period – speak of waves of mass migration, over land and sea, and warfare with mysterious invaders from afar. The Sumerians built a 100-kilometre wall to keep out climate refugees, which failed. (Some of these refugees settled further north, building a city called Babylon and seeding a new civilization.) Within decades the entire Bronze Age world had collapsed. The beneficiaries of this collapse were the migratory plains people.

The world's vast plains, with their constant winds and dry soils, make agriculture difficult but provide excellent pasture for horses and other herbivores. Thus they have been used by nomadic pastoralists and hunters for millennia. Many of these people also became cele-brated traders, exchanging their animal products for grown foods and other resources – or raiding from settled peoples with stored goods. This was the migration strategy that enabled our species to inhabit some of Earth's most difficult landscapes, and spread genes, cultures and resources across continents.

The Yamnaya steppes people were perhaps the most remarkable of these migratory cultures. They were the first to domesticate horses, and used them to conquer and colonize Europe some 5,000 years ago, becoming the third and final migratory people to change the DNA of all Europeans. The Yamnaya would have been an extraordinary sight, like nothing the indigenous European crop farmers had seen before: fair-skinned, dark-eyed warriors decorated with bronze jewellery, hur-tling along on horseback and with horse-drawn wagons. Their advanced metalwork and intricately patterned earthenware have been dug up from Scotland to Morocco. They also brought their Indo-European language, whose roots lay in northern Iran, and set up the first Eurasian trade in marijuana. The trade routes created by Yamnaya and their neighbours became part of the Silk Road several millennia later.

The Yamnaya were so transformative partly because they were networked – a web of mobile societies that formed an intercontinen-tal communication system – and because of their trading prowess. Successful migration relies on networks of collaboration and exchange. They were surely helped by timing, arriving in Europe shortly after plagues had devastated the continent. The Yamnaya migration was a violent invasion by any standards. Their bands stormed through Europe, overwhelming the indigenous population with sophisticated weaponry including battle-axes and a new, nimble bow-and-arrow. Around 90 per cent of the original gene pool was wiped out by the

Yamnaya, including all of the men in what is now Spain and Portugal.

Within a couple of centuries, the Yamnaya migration had revolutionized European society, culture and genes, ushering in the Bronze Age. Today, most people in Europe have light skin and half the world's people speak an Indo-European language. At least 70 per cent of European DNA is inherited from Anatolian migrants who arrived either 8,000 years ago as farmers, or 5,000 years ago via the steppes; the rest is hunter-gatherer DNA, which previously dominated. These were the last significant changes to European DNA, but it wouldn't be the last cultural change.

Over the following millennia, nomadic steppes warriors continually plundered agricultural settlements and cities, successfully occupying the sparse plains by parasitizing on the agriculturalists, but they were continually forced to retreat to where they could feed their horses – or to become settled themselves: this was the origin of the ancient Greeks and the Ottoman Turks, for example. These agile warriors brought down entire empires and, again, left their mark on the gene pool: Genghis Khan's DNA is present in around 1 in 200 men alive today, some 16 million people.

The oceans, too, hosted migratory raiders, including the Philistines (the 'Sea-Peoples'), who repeatedly attacked Egypt and Canaan (roughly modern-day Israel) where they went on to settle, founding Palestine; and the Vikings, who also regularly parasitized from settled communities, riding agile warrior vessels that made surprise attacks on coastal populations. 'Viking' literally means 'to raid' and some of their raids were carried out in mad furies by Vikings known as berserkers. As the thus-assailed intellectual Photios of Constantinople bewailed in 860: 'Why is this thick, sudden hailstorm of barbarians burst forth?'

Most of these nomadic groups disappeared or settled through agricultural expansion, although pastoralists still roam the grasslands, including the Mongolian steppes, Maasai Mara and Patagonia. Theirs is a far more sustainable use of marginal or grass lands.

LONG-DISTANCE VOYAGES

Human migration has also been driven by the quest to discover new lands and resources. Our curiosity and intrepid spirit has led people to explore beyond the safety net of their known world, and into the deepest seas, around the poles and to outer space. Migrations have distributed people across the Earth's landscapes, spreading our genes, cultural practices, beliefs and technologies.

For thousands of years, Polynesians used migration to deal with overpopulation and civil wars in their restricted island environments. Polynesian wayfinders were able to expertly navigate across thousands of kilometres of open ocean using the stars and their unparalleled knowledge of ocean currents, occupying remote islands from Hawaii to Indonesia.

By contrast, European navigators discovering New Worlds (and new parts of Old Worlds) often found them already peopled. The consequence of these migrations was rarely benign, and the navigators invariably stole resources, lands and people from one place, to migrate them elsewhere. The immigrants often asserted their own culture over the native culture in the belief that it was inherently better, seeding the racist idea that some people were 'primitive'. These journeys brought new people into contact, globalizing humanity and our operations, and advancing our knowledge of the world. It also brought death and disease, devastated established cultures, and dramatically altered the environment. A power imbalance still persists for descendants of colonizers over colonized, evident in lasting socioeconomic inequalities.

Migration drove the industrialized world, and a part of it was in pursuit of forced human labour to transform the environment and the wealth of nations. Slavery is an ancient practice but the transatlantic slave trade industrialized the enterprise to a horrific scale: some 12 million Africans were shipped to the Americas over 400 years. The legacy of such dramatic movement of, principally, young men from one society to another has had lasting affects on genes, culture, society and demography. Although Africans had been present in Europe and Asia for millennia, their numbers were relatively small, and the populations of Europeans and Asians migrating to the heart of Africa had also been insignificant. This is partly down to Africa's geography of desert, dense jungle and unnavigable rivers. Climate and ecology also played a role – the farming methods used in Eurasia are not suitable for the tropics, and African diseases proved deadly to many would-be colonizers. So this part of the world had been excluded from the large-scale intermixing of Eurasia until the slave trade. Much of the genetic and cultural diversity that we enjoy in the Americas and beyond has a brutal, shameful history, achieved through great suffering.

The great mixture of genes, people, cultures and technologies that exists today owes much of its variety and complexity to the human desire to migrate. The benefits of trade forced us to cooperate with other tribes with different social norms, genes and technologies. It thus expanded our social networks and our society's collective knowledge, and encouraged us to explore our environment in search of valuable raw materials. Sometimes cultural practices migrate with resources between populations; sometimes the people themselves migrate and integrate, bringing their technologies with them. The new advances in population genetics, archaeology, palaeontology and linguistics are starting to give us a much fuller picture of our migratory legacy. For example, we can see where Anglo-Saxon migrants settled in Britain, down to individual towns, because they changed the gene pool there; the Romans, Vikings and Normans, whose invasions transformed British cultural history, left much smaller legacies in our living DNA records.

Migration helped spread advantageous genetic modifications throughout humanity, often mixing and recombining them into new variants. The Yamnaya brought with them the genes to digest milk into adulthood – an adaptation they acquired through herding. This extra source of calories would have been of great advantage to stunted and malnourished crop farmers. The Yamnaya genes for fair skin, too, would have benefited the dark-skinned native farmers who had little access to food sources of vitamin D, especially during the short days of the northern winters. In small populations, even a slight advantage can lead to a gene being proliferated.

The past couple of centuries have seen the biggest mixture of peoples, generating a truly globalized world, as vast populations crossed continents, fleeing conflict or seeking a better life. In some cases, immigration was state-led, to fulfil a labour requirement. This is what enabled the rapid development of Australia, California and the British National Health Service. It is also what draws migrant workers to Dubai, students to university towns and scientists to international collaborations such as CERN. Indeed it is no coincidence that scientific discovery and progress arises out of collaborative labs of migrants – they are necessary to fertilize innovation, consolidate learning and diversify ideas.

Migration, therefore, is not simply a feature of human societies: it may be essential. Without sufficient migration, cultural and genetic complexity can drop to the extent that societies struggle to survive, or even go extinct. This, for instance, is what happened to a small detachment of Canadian Palaeo-Eskimos, who navigated their way across the hostile ice and sea from Siberia to North America and down through Canada, around 6,000 years ago. Despite sharing territory with highly sophisticated indigenous Amerindian populations in southern Canada, they deliberately isolated themselves. Over time, the Eskimos struggled and went extinct. Their health may have been poorer because of inbreeding, and their culture regressed. An entire people disappeared through lack of communication with others.

Social networks create synergy, allowing configured groups to achieve things that a disconnected collection of people could not. Migration relies on networks of cooperation to bridge difficult problems, just as it does to bridge alien and inhospitable lands. Pioneers then provide the arrival network for subsequent migrants from the same origin community. As a result migration isn't random around the globe but follows the inherited routes and channels carved out by our many historical predecessors, from the Crusaders to the Silk Road merchants to the post-colonial diaspora.

Migration, then, has continued throughout human history, sometimes involving large populations, sometimes just a few people, and often moving into territory occupied by existing societies. Entering another tribe's territory is dangerous, and in normalizing this our ancestors broke from primate behaviour – chimps, for instance, attack any intruders, often killing them. Most human societies, by contrast, have social norms that welcome strangers, and the reputation of the group and the prestige of its leaders hangs on ensuring visitors are treated hospitably.

We also make the most of our family connections. One reason human inter-group interactions are often cooperative rather than hostile is because of the far greater benefits of trading between groups; the other is our relatedness. Our extended families, which include in-laws, often straddle group boundaries, so most people are closely related to people who live in different districts, countries or even continents, and most people on Earth speak multiple languages for this reason. Beyond our immediate family, we are surprisingly closely related to each other – by which I mean, humans are all genetically very similar in comparison to the diversity within other species: two humans picked at random differ in their DNA by an average of 0.1 per cent, much less than two random chimpanzees. The clusters of genetic differences between populations, even across continents, such as between a Sri Lankan and a Swede, are tiny anomalies. This is partly owing to ancestral population crashes, but mainly because of our migratory interbreeding, facilitated through trade networks. Sub-Saharan Africa, the birthplace of our species, is home to one-eighth of the world's population and the greatest genetic diversity – the difference in DNA between people in West Africa and East Africa is double that between Europeans and East Asians. Yet despite our hundreds of thousands of years' worth of evolved complexity, societies regularly define people as being either 'Black' or 'White', as though skin tone is a binary state and can reveal any meaningful distinction about a person's genetic 'purity' or 'race'.

The dense interconnectedness of the human family, our genetic similarity, means that, in terms of our biology, there are no different races. We can all claim ancestry across the world. The ties that bind us to a particular land – our national identity – are cultural and based on an arbitrary moment in time, usually the location on Earth of your mother at your birth. Genetic features overlap and are dispersed across cultural and geographical boundaries, so that genetic variation within populations is at least as great as between them. (Even where strict cultural norms forbade intermarriage between groups, the genetic evidence reveals that it continued.)

Great migrations of people, invading, fleeing, crusading, exploring, roaming, colonizing, slave-trading; uprooted for war, work, or fortune – all have contributed to the genetic mixing of the past millennium, particularly in recent centuries. As the branches of our human family tree become ever more entwined, this is leading towards a situation where populations feature such widespread mixed heritage that tribal in-group/out-group prejudice will no longer be possible based on visible distinctions[7] – the great climate migration could hasten this to within a few generations. The genetic difference between people of Europe and West Asia has more than halved over the past ten millennia, for instance. In other words, describing differences between people based on the fallacy of biological 'race' will no longer be credible.

The focal points for cross-cultural trade networks are our cities. And cities cannot exist in isolation: they rely on trade networks of merchants, diplomats and artisans importing new resources and ideas. The city dweller has a very different social network to the villager, and a different sense of identity and of their ties to the land. In comparison to villagers, city people are more likely to know a diverse range of people who are less likely to know each other, and that expansion of networks increases the potential for fruitful interactions and drives innovation. Cities, then, are cultural factories, attracting diverse, concentrated populations and, like all social networks, they are synergistic: increase the population of a city by 100 per cent and innovation increases by 115 per cent.

Historically, people moving for urban opportunities accelerated technological progress, generating civilizations, writing and modern industrial economies. City living also changed our health for the worse with the emergence of new epidemics that dramatically increased mortality – indeed, until the twentieth century, cities were so dangerous that their populations were only sustained through constant immigration from rural areas. It is only with the advent of sanitation, sewerage and modern medicine that cities became comparatively safer places to live.

The genetic impacts of cities can outlive the cities themselves. Some 400 years ago, a charismatic leader of West Africa's Kuba tribe, called

Shyaam a-Mbul, formed a kingdom in what is now the central-southwest of the Democratic Republic of Congo that drew together the region's many ethnic tribes into a large, sophisticated city-state. The Kuba Kingdom had an incredibly modern political system, including a constitution, elected political offices, trial by jury and the provision of public goods and social support. It became a wealthy hub of innovation, famous for its artworks. Belgian colonization at the end of the nineteenth century did much to diminish this remarkable cosmopolitan state, but its legacy lives on: populations with Kuba ancestry are far more genetically diverse than others from the area.

All of the world's major cities were created by migration, often by refugees, as Europe's Rome and Venice. The equivalent of a million-person city will be built every ten days over the next eighty years, as we undergo the biggest mass migration in our history. Most urbanization in the coming decades will consist of poor people in Africa and Asia migrating from rural areas for paid work. Almost all of them will live in slums at densities as great as 2,500 people per hectare, sharing just two or three toilets (the same number as the average US family home). There are around thirty megacities today, and by 2050 they are expected to merge into dozens of mega-regions, like Hong Kong–Shenzhen–Guangzhou in China, where more than 100 million people will live in a seemingly endless city. Such mega-regions will probably become more influential than nation states.

We already see more power being devolved to cities, which now have considerable agency over issues from migration to climate action. For instance, nation states supposedly control the flow of refugees and immigrants into a country, but, in reality, it's often city authorities that do. It's a city's government that decides on issues such as housing and employment for its occupants, whether they were born there or entered legally or as undocumented migrants; and many have begun issuing 'urban visas' – New York's NYCID programme, for instance, allows anyone with proof of residency in the city (like an electricity bill) to get a government-issued identification card. According to Benjamin Barber, founder of the Global Parliament of Mayors: 'The city in effect becomes the one that authorizes, controls, registers and oversees the immigrant population that has gotten into these places without the permission of the nation states, which aren't in a position to control them and don't.'[8]

We are still in transition and cities are sophisticated organisms: the bigger they grow, the faster they operate and innovate.

The story of human migration is the story of our genes, culture and landscape as they have changed over millennia. It is the story of nomadism and shifting agriculture, of the endless tussle between the peoples of the plains and the soil, of empires that expanded and vanished, explorers that reached the furthest corners of our planet and the people that followed in their footsteps. It is a story of belonging and its opposite: the displaced, the homeless and the stateless. And it is the story of our most triumphant creation of a human niche, the city, and the billions of migrants cities attract.

We spread across the world through networks that we ourselves maintain – where they are dense and well-connected, movement is easy and our societies thrive; where networks are disrupted, movement is restricted, and societies and cultures decline. Even if we do not ourselves migrate, our ancestors did, and we are utterly dependent on the migration of other people and resources for the modern world to function. The exchange of people and goods from other places, the words we speak, the foods we eat, the music we listen to ... it all relies on our mobilized human world.

Today, though, there have never been more barriers to migration, as countries seal borders and build walls. As humanity faces its greatest environmental challenge – a population of 10 billion people, resource limitations, and a demographic crisis – we should not be handicapping ourselves by limiting our most important survival tool. We will only meet our global challenges through planned and extensive human movement and redistribution. However, as we've seen, large-scale migration has been bloody and brutal, and in the technologically advanced world we live in today the possibilities for catastrophe are ever-present. This time we need a globally managed effort that recognizes our shared humanity on this shared earth – we need lawful, safe, planned and facilitated migration.

4

Bordering on Insanity

In the year 1800, humans reached a global population of 1 billion. It had taken us around 300,000 years. Just 200 years later, our population reached 6 billion. Twenty years after that, in 2020, there were almost 8 billion of us. We are by far the most populous big animal, an evolutionary success owed, in large part, to our ancestors' migrations.

We live everywhere and yet we are not dispersed evenly across our globe. Instead, the world's human population is concentrated in clusters: some relatively small areas are very densely populated, whereas others are virtually empty. Bangladesh, for instance, is occupied by 1,252 people per square kilometre; this is almost three times as dense as its neighbour, India, and more than 400 times as dense as Australia, where there are just three people per square kilometre.

An interplanetary visitor might find this state of affairs surprising, and assume perhaps that Bangladesh was home to a large proportion of the world's food or some other highly desirable resource. If you look closer at Earth, though, you'll see that actually most people are clustered in even tinier areas: the cities. Around 42,000 people per square kilometre in Manila, for instance. Or what about Dharavi, a slum in Mumbai, which has a population of a million people living in just 2 square kilometres.

This isn't necessarily a bad thing: it makes sense for people to go to where they and their children have the best opportunity for a good life, where they have enough to eat, a safe place to live and the ability to study or earn. Cities can offer unparalleled rewards in that respect. However, if we were to map those 'best opportunities', we find that the places where people have the best chance of a good life are not a match for the map of places where the most people live. This would surely confuse our interplanetary visitor – after all, other species live in the environments they are most suitable for: they have evolved to be particularly adapted for their specific niche. Humans, though, have created a problem for themselves.

Most of the world's population is clustered around the 27th parallel, which has traditionally been the latitude of most comfortable climate and fertile land, but this is changing. Climate adaptation will mean chasing our shifting niche as it migrates north. On average, climate niches around the world are moving polewards at a pace of 115 centimetres per day, although it's far faster in some places.[1] Ecologists calculate today's velocity of climate change at 0.42 kilometres per year, which means that's the speed species, including humans, need to migrate away from the equator to keep enjoying the same climate conditions.

We have no choice about the place we were born, we arrived there through our ancestors' migration, however recent or ancient. Many people today live in places that are particularly vulnerable to environmental catastrophe, overcrowding and poverty. This is going to get immeasurably worse. Relocation has been our solution in the past, and could solve many of these problems today – and not just for the migrants. Host nations and the migrants' origin nations also win out in multiple ways, not least economically.

The human climate niche, home to billions of people

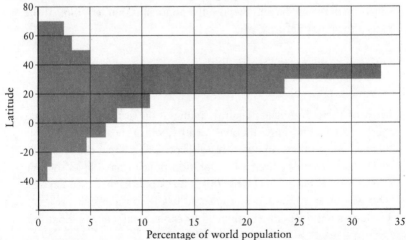

The problem is, people in this situation face enormous difficulties leaving and moving somewhere safer. As we've seen, the only reason that we humans can be so flexible about our 'niche' that we are able to live in any environment, is because we are supported by the life-giving social networks that enable this. Leaving this support system is daunting, particularly if you risk not being included in the networks of your destination. It can also be impossible.

The main barrier is our system of borders – movement restrictions either imposed by someone's own state or by the states they wish to enter. While in the late nineteenth century 14 per cent of the global population were international migrants, today just over 3 per cent (though of a much larger population) are. However, migrants contribute around 10 per cent of global GDP or $6.7 trillion – some $3 trillion more than they would have produced in their origin countries.[2] Several economists calculate that if borders were removed, global GDP would soar by between 100 and 150 per cent – increasing, in other words, by at least $90 trillion a year.[3] As we shall see, managed well, migration benefits everyone.

This might seem surprising. Surely migrants are a burden on our already overstretched social welfare systems, which we pay for out of our taxes, adding the sting of unfairness to it? Why should jobs and other opportunities go to migrants rather than those who deserve them by being born here? These arguments have been made very persuasively in the media and by populist politicians, but let's look at the facts – they may surprise you.

WHO BELONGS HERE?

Most of these arguments against immigration rest on the idea that there is a true and pure national identity, which means some people 'belong' while others do not. It is no surprise to me that the architects of Donald Trump's extreme immigration policies have links to 'race science' and eugenics.[4] As I write this, a Black politician, David Lammy, who was born in London in 1972 and has lived there since, is being asked on a national radio station, 'How can you call yourself English?'

The caller says that her own ancestors go back to Anglo-Saxon times, whereas his are clearly African-Caribbean. There's a lot to unpack here about heritage, colonialism, national identity and more. (It's worth noting that of the thousand or so people who held their lands directly from the king in 1086, the Domesday Book tells us that only thirteen were English. All the rest were recent immigrants.) But at its heart this is a case of one person claiming that someone with more melanated skin cannot be a member of her 'white' tribe, and also that only her tribe *belongs* here – is the rightful owner-occupier of the land. It's easy to simply dismiss this as stupid racism, but it helps to understand that there are long evolutionary roots to prejudice.

Humans can be easily suspicious and mistrustful of outsiders, so it doesn't take much to persuade us that people from elsewhere are not as worthy or deserving of our resources and, today, in many countries, there is no shortage of those trying to do so. Being able to rely on each other in our social group is so important to human survival that we have evolved countless ways to prove that we belong to our 'tribe' and are loyal and reliable, and therefore deserving of its security and other benefits. This is important because humans – unlike other social animals like bees or ants – rely on social groups that are not exclusively family members who share their genes. From birth onward we learn consciously and subconsciously the social norms – the behaviours and cultural practices – of our tribe, effortlessly 'belonging' to that culture. The more you share someone's social norms, the better you can predict their behaviour, so the easier it is to decide whether you can trust them to act in your interests – which lowers the costs of exchanges and interactions between people. Since this tribal allegiance is predicated on being an insider, it means there must be outsiders, from whom we must protect our resources, who might want to take advantage of us, and whom we can't necessarily trust.

Our out-group prejudice is learned in early childhood, and although we often dress up our animosity in terms of objections to cultural differences rather than the individuals themselves, in reality these are deeply held cognitive patterns. People feel a connection to other members of their group: their brains respond in empathy at their pain, for instance. But if they are told that another person is a member of an out-group, such as a rival sports fan, they stop empathizing. By identifying people as members of an out-group, we clarify the parameters of the in-group and make our own position more secure within it.

A consequence of individual identity being so bound with the group's is that if someone switches tribes, she risks losing her own sense of identity and feeling alienated by both tribes, which has mental health implications (migrants have a higher incidence of schizophrenia, for instance, than non-migrants). But people will continue to try because being in-group is so valuable for the protections and other benefits it brings.

The more similar people look, and the more similar their cultural background, the more important become the identifying tokens and social norms. Catholics and Protestants in Northern Ireland, like Hutus and Tutsis in Rwanda, look and sound the same as each other, and so they have had to accentuate any small differences, whether they be rituals, religions or foods. We craft a group identity through stories that cast us as the righteous side – as heroes or unfairly treated victims – in competitions with other groups. These compelling narratives can also be an extremely effective way of getting socially alike individuals to kill each other on the basis of belonging to opposed groups.

It is when a group feels threatened that it binds itself most strongly in defence of its collective tribal interests – even five-year-old children act more cooperatively and generously with their group when it is under threat. Groups of men who fight together are more likely to survive, and as generals know, each soldier is more likely to survive when the whole troop is prepared to die for each other.

This also offers politicians a cynical way to strengthen national institutions and keep politically diverse societies cohesive: via competition and conflict with another group. It helps explain the rise of nationalism: its manifestation is indicative of a threat to the group, and this works as a feedback loop, convincing the group it is under threat from migrants or neighbouring countries. However, the threat in most of these nations today is not external – rather, the threat is from internal social divisions and inequalities, between rich and poor; old and young; rural and urban; university educated and not. It's important to recognize that although the capacity for tribalism is there in all of us and has manifested throughout our history, it is not inevitable and is mediated by social norms that can be inclusive or exclusive. Discrimination and oppression are not the inevitable results of migration.

The great paradox of human culture is that while we are primed for tribalism, we rely on networks of cooperation between our tribes to exchange ideas, resources and genes, as we've seen. We are skilled at welcoming strangers into our groups with well-honed social strategies to enable cooperation between groups as well as within them. Most people on Earth are multilingual, and many of us share family – if not our parents, then cousins – between multiple groups. Our networks span multiple groups and each node – each person – within our own social network will have their own broad network. We are all connected, then, within a few degrees of separation. David Lammy, for instance, is English by birth, has Guyanese-born parents, and his ancestors include Africans who were enslaved for centuries by the Dutch and British in what is now Guyana and Barbados. He also has Scottish ancestry, very likely from the rape of one of these slaves, and thus he is descended from slaves and slave owners. His parents were recruited from their British colony to the Mother Country, as British subjects, after the Second World War to help rebuild the nation. He describes his identity variously as British, English, a Londoner, European, and of African-Caribbean heritage. We all have similar entangled ancestry within a few generations, even if it doesn't immediately reveal itself in our skin tone.

Although tribalism can cause difficulties for migrants, there's evidence that it may actually be declining, especially among younger people and in melting-pot cities, where it is difficult to produce the 'pure' in-group by which to define an out-group. In the UK, despite Brexit – or perhaps because of it – concern about immigration has been falling. Today it is at its lowest level this century, according to surveys by Ipsos.

Millennials everywhere are much less likely to view nationality in racial terms. Generally, they value mobility and are more concerned about rising living costs than patriotism (especially as defined by Bertrand Russell, as the willingness to kill or be killed for trivial reasons). Nationalist leadership, ironically, generally works to drive young citizens to leave their nation. We should bear in mind the preferences of younger generations, since it is their future we create today.

Concern about immigration at lowest level this century

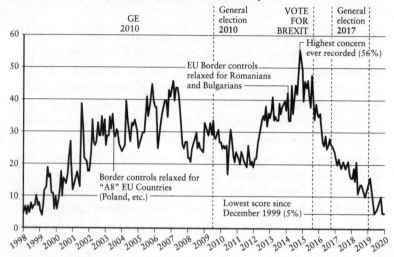

GE 2010

General election 2010

VOTE FOR BREXIT

General election 2017

EU Border controls relaxed for Romanians and Bulgarians

Highest concern ever recorded (56%)

Border controls relaxed for "A8" EU Countries (Poland, etc.)

Lowest score since December 1999 (5%)

Post the Brexit culture war, Britons remain positive about immigration

Q: On a scale of 0-10, has migration had a positive or negative impact on Britain?
(0 is 'very negative' and 10 is 'very positive')

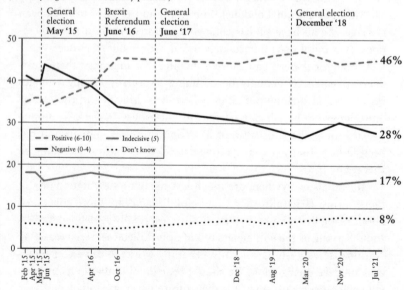

General election May '15

Brexit Referendum June '16

General election June '17

General election December '18

- - - Positive (6-10)   —— Indecisive (5)
—— Negative (0-4)   ••• Don't know

46%

28%

17%

8%

INVENTING THE NATION

The idea of keeping foreign people out using borders is relatively recent. States used to be far more concerned about stopping people from leaving than preventing their arrival. They needed their labour and taxes. Laws in Roman times, for instance, kept peasants and labourers bound to their farms. China, too, required 'zhuan' documents for any movement around the country. The same was true in Europe in medieval times. In the 1600s, for instance, English labourers needed locally issued passes to travel for work, partly to stop them 'benefits shopping' for parish poor relief. Similarly, in the medieval Islamic Caliphate, people needed to show their bara'a, a receipt of paid taxes, in order to travel to other parts of the Caliphate. A culture of discouraging people from leaving the country persisted into modern times – in 1816, a *Times* editorial described people who wanted to emigrate from Britain as either 'paupers and fools' or 'malignant outcasts . . . execrably base in their natures'.

Passports, as far as they existed, were letters requesting safe passage and conduct through a foreign territory and didn't restrict entry, although, during some periods, local authorities keen to extract payment would list city gates (*portes*) that the traveller was permitted to *pass* through. Sea ports, for instance, didn't require documents, as they were considered open trading centres.

The main reason for this was that an individual's nationality had little political meaning before the end of the eighteenth century. People had ethnic and cultural identities, but these didn't really define the political entity they lived in.

Human societal structures evolved in complexity as our populations grew from bands of hunter-gatherers to larger settled villages that were loosely networked. These alliances helped people to survive hardship and feed and defend themselves, but most of us can only keep track of social interactions with around 150 individuals – the so-called Dunbar number. Anthropologist Robin Dunbar discovered a ratio between primate brain size and the number of individuals in their group – which corresponds, he theorized, to a cognitive limit for social complexity. The human number of 150 turns out to be remarkably consistent across an array of social groupings, from hunter-gatherer societies to twentieth-century Christmas card lists. The way our societies grew past this Dunbar number was through the mechanism of hierarchies. Several villages banded together under one chief; several chiefdoms were then banded under a higher chief, and so on. To grow, more villages could be added and, if necessary, further layers of hierarchy.

This meant that leaders could coordinate large groups without anyone needing to keep personal track of more than 150 people. For instance, a chief would keep company with those in their immediate circle, as well as one chief in the hierarchical level above them, and several people from the levels below them. This extra social complexity enabled greater collective action. Villages became organized around market towns, grouped for tax purposes, to supply armies, or labour for harvesting or building infrastructure. From these hierarchies grew cities and empires – but not nation states. That's because almost the whole population, until recently, were farming peasants living with the real and frequent risk of starvation. As a result, people were largely self-organizing, and leaders did little actual governing. Mainly leaders were involved in fighting either to acquire more territory or to keep those territories they had.

Even relatively recently, rulers spent little time on domestic governance. In the eighteenth century, the Dutch and Swiss didn't even have a central government. Indeed, in the twenty-first century for nearly two years, neither did Belgium. Eastern European immigrants arriving in the US in the nineteenth century could say what village they came from but not often which country: it didn't matter to them. People defined themselves 'vertically' by who their rulers were. The land you lived on – and so you, for identity purposes – became the property of whichever ruler had acquired it though conquest, inheritance or marriage. There was little interaction between villages beyond the local market, so it was irrelevant to them whether the same king ruled over those people, or whether they shared interests. Anyway, alliances and territories were vague and changeable and often came under different jurisdictions for different purposes. Until the nineteenth century, even Britain had several different dialects and languages.

This loose system of governance limited the complexity of collective actions that could be scaled up by leaders to things like growing food, collecting taxes, fighting battles and keeping order. Some, such as the Roman Empire, were particularly successful at this and managed more, but generally, pre-modern societies were limited by the amount of energy they could harness, which was largely human and animal labour. When water power was rolled out during the Middle Ages, it led to an increase in production and trade, which boosted social complexity to the extent that decentralized feudal systems ceded to centralized monarchies that were almost constantly at war with each other. But these were not nation states.

The seeds of change occurred in 1648, when two peace treaties were signed in northern Germany, ending centuries of war that had killed millions, including, most recently, the Thirty Years War. Europe's Peace of Westphalia essentially declared existing kingdoms, empires and other political entities 'sovereign': none was to interfere in the domestic affairs of the others. However, sovereign states were defined by the family trees of their leaders, not by their people's national identities. The word 'international' had no meaning, and doesn't appear until the end of the eighteenth century. By then, coal power had enabled production on an industrial scale, and the associated ramping up of social complexity enabled far greater complexity of government and collective activity. And it demanded a new type of government.

It took revolutions to create the first nation states defined by the national identity of their citizens rather than the bloodlines of their rulers. They were pioneered in Latin America by the Creoles, descendants of European colonizers who wanted independence from Spanish control (and to raise their own social standing). In Europe, the nation state was created by French revolutionaries. In 1800 almost nobody in France thought of themselves as French, and only about 10 per cent could even speak the language. By 1900 they all did.[5] However, as recently as 1940, Winston Churchill proposed full political union with France: 'France and Great Britain shall no longer be two nations, but one Franco-British Union.' The French rejected this.

After the First World War, with the end of the Habsburgs and other multinational empires in Europe, state borders were redrawn along linguistic and cultural lines and the nation state became the norm. Much of this was pragmatic. With the move from agricultural economies to industrialization, micro-states became less viable because they didn't always contain the necessary resources like coal or steel. On the other hand, large empires became harder to manage because they needed more governance. Nation states, then, were economically the most efficient. It's worth noting that most of the world's 200 nations, which we consider immemorially set in stone, were created when the global population was less than one-quarter of today's.

They may have been created politically, but they still had to be invented. In 1860, when Italy unified, only 2.5 per cent of its citizens actually spoke Italian; even the leaders spoke French to each other – one famously said that, having created Italy, they now had to create Italians.

Nation building required the creation of an ideology of nationalism that emotionally equated the nation with people's Dunbar circle of friends and family. Instead of the vertical state, it created a horizontal state. National identity was deliberately fostered with mass education and mass media, including newspapers and other literature that standardized vernaculars and created a horizontal linguistic community of people that read and cared about the same things. Once people's nationality became important, it paved the way for identity papers and the modern state.

BUREAUCRACY IS THE SECRET INGREDIENT OF PATRIOTISM

You may think that flags, anthems and an army to guard your territory is what's needed to develop a sense of nationalism, but it's more accurately the success of its bureaucracy. Greater government intervention in people's lives and the creation of a broad systemic bureaucracy were needed to run a complex industrial society and they also forged national identity in its citizens. For instance, Prussia began paying unemployment benefit in the 1880s, which was issued initially in a worker's home village, where people and their circumstances were known. But it was also paid to people where they migrated for work, which meant a new layer of bureaucracy to establish who was Prussian and therefore entitled to benefits. This resulted in citizenship papers and controlled borders. As governments exerted greater control, people got more state benefits from their taxes, and more rights, such as voting, which engendered a feeling of ownership over the state. It became their nation.

Nation states, then, are an unnatural, artificial social structure that emerged out of the complexity of the industrial revolution, and they are predicated on the mythology that the world is made of distinct, homogeneous groups that occupy separate portions of the globe, and claim most people's primary allegiance. The reality is far messier, as we've seen. Most people speak the languages of multiple groups, and ethnic and cultural pluralism is the norm. A person may identify with the religion of one 'tribe', the cuisine of another, and so too for fashion, language, cultural references and lifestyles of other overlapping or separate 'tribal' identities. People always have a sense of belonging to numerous groups, and the idea that a person's identity and well-being is primarily tied to that of some invented national group is a stretch, even if this is presupposed by many governments. The Irish political scientist Benedict Anderson famously described nation states as 'imagined communities'.

It is, then, hardly surprising that the nation-state model so often fails – there have been around 200 civil wars since 1960, with one-fifth of nations experiencing at least ten years of civil war.[6] Such failures, in which nations are split along sectarian lines, are often used to support the idea that nations should be forged from single homogeneous 'tribes' – disasters are blamed on the colonial legacy of collating multiple groups within arbitrarily created boundaries, for instance. However, there are plenty of examples of nation states that work well despite being made up of different 'tribes', such as Singapore, Malaysia and Tanzania, or nations created from global migrants like Australia, Canada and the US. And anyway, to some degree all nation states have been formed from a mixture of groups. The UAE, an extreme example, has no majority tribe; everyone living there is a minority. When nation states falter or fail, the problem is not diversity itself, but not enough official inclusiveness – equity in the eyes of the state, regardless of which other groups a person belongs to.

An insecure government allied to a specific group, which it favours over others, breeds discontent and pitches one group against others – this results in people falling back on trusted alliances based on kinship with other groups. By contrast, a democracy with a mandate of official inclusiveness from its people is generally more stable – but it needs underpinning by a complex bureaucracy; failed states are generally those without complex bureaucracies.[7] Bureaucracy creates functional complexity in society as well as being a result of a functional nation state. Bringing diverse groups together into a functional system needs a complex bureaucracy to work, but the diversity of the nation's people must be equally included by officialdom. Nations have navigated this in various ways, from ethnic cleansing to remove diversity (as is under way in Uighur Muslim communities in China) to devolving power to local communities, giving them voice and agency over their own affairs within the nation state (as is the case in Canada or for Switzerland's cantons). By embracing multiple groups, languages and cultures as equally legitimate, a country like Tanzania can get along as a national mosaic of at least 100 different ethnic groups and languages. In Singapore, which has consciously pursued an integrated multi-ethnic population, at least one-fifth of marriages are interracial, giving rise to a generation of 'Chindian' children, for instance.

Unjust hierarchies between groups make this harder, particularly when they were imposed on a majority by a minority, such as through colonization. There remains huge injustice between groups of people based on how they or their ancestors derive their national identity. In many cases, original inhabitants never received official citizenship – it wasn't until the 1960s that indigenous Australians were given citizenship of the land their ancestors discovered and inhabited for 60,000 years.

In April 2021, Governor Kristi Noem tweeted: 'South Dakota won't be taking any illegal immigrants that the Biden Administration wants to relocate. My message to illegal immigrants . . . call me when you're an American.'

Consider that South Dakota only exists because thousands of undocumented immigrants from Europe used the Homestead Act from 1860 to 1920 to steal land from Native Americans without compensation or reparations. This kind of exclusive attitude from a leader weakens the sense of shared citizenship among all, creating divisions between residents who are deemed to belong and those who are not.

Official inclusion by the national bureaucracy is a starting point for building national identity in all citizens, particularly with a large influx of migrants, but the legacy of decades or centuries of injustice persists socially, economically and politically. We do, after all, live in a time when it is possible to question whether a Member of Parliament, born in London and speaking clear, accentless English, can legitimately call themselves English based on the colour of their skin. Officially, however, Lammy has the same UK passport and the same rights as his interrogator.

HOW WE ENDED FREE MOVEMENT

After it had been introduced in France, the nation-state model spread, and so did passports. However, this soon became problematic: the explosive activity of the industrial revolution required free movement of labour, trade and money for production, so any passport requirements were relaxed. In 1872, the British foreign secretary, Earl Granville, wrote: 'all foreigners have the unrestricted right of entrance into and residence in this country.' Indeed, Britain had a proud history of granting asylum to people. Between 1823 and the Aliens Act of 1905, not a single foreign citizen was either refused entry or expelled from the country – an 1853 *Times* editorial declared: 'This country is an asylum nation, and it will defend the asylum to the last ounce of its treasure, and to the last drop of its blood.' Britain has moved some distance to today's 'hostile environment'.

Legal experts remained divided over whether states even had the right to control people's international movements, well into the twentieth century. Then the nationalism that propelled Europe towards war radically shifted attitudes towards suspicion of foreigners and stoked fears about whether they were spies. Passport controls were applied, and over the past century they've become more restrictive and globally enforced. Borders, by their very existence, are 'othering' structures, and nations have been quick to use them to maintain their in-group/ out-group hierarchies. For instance, border policies have included the United States' Chinese Exclusion Act (1882), the White Australia Policy (which lasted until 1973) and Britain's Commonwealth Immigrants Act of 1962 (which effectively reduced citizenship rights for darker-skinned people born outside the UK).

Globally, in the past few decades, borders have become more restrictive, the rhetoric around immigration has grown hostile, particularly to refugees, and some states are even paying others to imprison would-be migrants within their borders – the EU has been compensating Libya for preventing people from leaving its borders, for instance. (This strategy usually fails, with the exception of North Korea, where the leadership has been largely successful in enforcing it.) Denmark and the UK have been floating plans to send asylum-seekers to Rwanda for their claims to be processed; Australia banishes its claimants to detention centres in Papua New Guinea and Nauru, where they languish for years if they haven't died from violence, lack of medical care or suicide.

Migrants have been portrayed as a security threat, and pledging stronger borders is a vote-winner across the world. People are told that immigrants will steal their jobs, drive down wages, and freeload on social services, from hospital care to government benefits. Politicians have allowed the rhetoric around human movement to become truly toxic – in 2012, the UK introduced a border policy officially called 'hostile environment', which was widely criticized, and blamed for fostering xenophobia by the United Nations Human Rights Council. People are banned from crossing our invented lines of geography not because of what they have done but because of where or who they were born.

We are now nearly 8 billion people, locked into geographical position on this planet by chance of birth. Passports and privileges are bestowed or inherited unequally, enabling some people to explore our planet unhindered, whereas others are trapped. Wall-building and demonization of migrants results in death, slavery and hate crime. But it does not prevent migration. People will continue to move, and we should. Migration is inevitable; people have no choice. It must be facilitated.

BEYOND BORDERS

The frontline in Europe's war against migrants is the Mediterranean Sea, patrolled by Italian warships tasked with intercepting small EU-bound vessels and forcing them instead to ports in Libya on the North African coast. One such warship, the *Caprera*, was singled out for praise by Italy's anti-migrant interior minister for 'defending our security', after it intercepted more than eighty migrant boats, carrying more than 7,000 people. 'Honour!' he tweeted, posting a photo of himself with the crew in 2018.[8] However, during an inspection of the *Caprera* that same year, police discovered more than 700,000 contraband cigarettes and large numbers of other smuggled goods imported by the crew from Libya to be sold for profit in Italy. On further investigation, the smuggling enterprise turned out to be even greater, a highly lucrative business involving several other military ships. 'I felt like Dante descending into the inferno,' said Lieutenant-Colonel Gabriele Gargano, the police officer who led the investigation.

The case highlights a central absurdity around today's attitude to migration. Immigration controls are regarded as essential, but these border restrictions are for people, not stuff. Huge effort goes into enabling the cross-border migration of goods, services and money. It's big business: every year more than 11 billion tonnes of stuff is shipped around the world – the equivalent of 1.5 tonnes per person per year. Whereas humans, who are a key part of all this economic activity, are unable to move freely. Industrialized nations with big demographic challenges and important labour shortages are blocked from employing migrants who are desperate for jobs.

The tragedy is that the African migrants had paid vast amounts in 'import duty' to smugglers, yet never made it safely across. Several drowned through these military interventions. Without safe, legal routes for migration, states miss out on this immigration 'tax' and the wealth of other benefits migrants bring. And, of course, migrants lose out on the opportunity to build new lives in safety and stability.

MIGRATION IS AN ECONOMIC –
NOT A SECURITY – ISSUE

Currently, there is no global body or organization overseeing the movement of people worldwide. Governments belong to the International Organization for Migration, but this is an independent, 'related organization' of the United Nations, rather than an actual UN agency: it is not subject to the direct oversight of the General

Assembly and cannot set common policy that would enable countries to capitalize on the opportunities immigrants offer. Migrants are usually managed by each individual nation's foreign ministry, rather than the labour ministry, so decisions are made without the information or coordinated policies to match people with job markets. We need a new mechanism to manage global labour mobility far more effectively and efficiently – it is our biggest economic resource, after all. It's nonsensical to have wide-ranging global trade deals for the movement of other resources and products while stymieing labour movement.

In July 2018, the UN Global Compact for Safe, Orderly and Regular Migration was agreed by all 193 members, except the United States. But at the ceremony to adopt the text that December, only 164 countries formally adopted it, with refusers including Hungary, Austria, Italy, Poland, Slovakia, Chile and Australia. The agreement, which is not legally binding, emphasizes that all migrants are entitled to universal human rights, and aspires to eliminate all forms of discrimination against migrants and their families. The document is a statement of intent and 'reaffirms the sovereign right of states to determine their national migration policy'. It is, then, weak and not fit for purpose.

Up to 1.5 billion people will have to leave their homes by 2050, according to the International Organization for Migration, and a recent analysis by a different team of scientists pushes the figure up to 3 billion by 2070.[9] Most of the world's displaced people already come from the global south, from the tropics and the places most affected by climate change. Many of these places are experiencing population growth, especially in Africa, with increasing numbers of young people forced to migrate for safety or new opportunities. For wealthy nations, migration solves many of the issues of workforce shortages. Consider that sub-Saharan Africa alone will have a resource of 800 million new working-age people in the next three decades. India already has more millennials than make up the entire population of the US or EU, as does China. Some 60 per cent of the world's population is under the age of forty, half of these (and growing) under twenty, and they will form most of the world's people for the rest of this century. Many of these young, energetic jobseekers are going to move as the world heats – will they add to economic growth or will their talents be wasted?

The conversation about migration has become stuck on what ought to be allowed, rather than planning for what will occur. Nations need to move on from the idea of *controlling* migration to *managing* migration. At the very least, we need new mechanisms for lawful economic labour migration and mobility, and far better protection for those fleeing danger. Within days of Russia's invasion of Ukraine in 2022, EU leaders enacted an open-border policy for refugees fleeing the conflict, giving them the right to live and work across the bloc for three years, and helping with housing, education, transport and other needs. The policy undoubtedly saved lives but additionally, by not requiring millions of people to go through protracted asylum processes, enabled the refugees to disperse to places where they could better help themselves and be helped by local communities. Across the EU, people came together in their communities, on social media, and through institutions to organize ways of hosting refugees. They offered rooms in their homes, collected donations of clothes and toys, set up language camps and mental health support – all of which was legal because of the open-border policy. This reduced the burden for central government, host towns and refugees alike.

Wherever migrants settle, they create active markets, but policies today limit the scope and potential of these markets.

Migration requires funds, contacts and courage. It usually involves a degree of hardship, at least initially, as people are wrenched from their families, familiar surroundings and language; often it involves uncertainty over basic food and shelter. Some countries make it almost impossible to move for work, and in others, parents are forced to leave behind children that they may never see grow up – an entire generation of Chinese children has reached adulthood seeing their parents only for a week or so once a year during spring festival. Three times as many people migrate within their country as internationally, and China is home to more internal migrants than there are migrants elsewhere. In China, hundreds of millions of people are caught in limbo between the village and cities, unable to fully transition due to archaic land laws and the lack of social housing, childcare, schools or other public facilities in the cities. The villages are sustained through remittances from absent workers, who cannot sell their farms for fear of losing their land, which is their only social security. Left-behind, isolated children then become primary caregivers for their ageing relatives while only teenagers themselves. Migrant workers cannot afford to buy homes in the city and so return to the village on retirement, restarting the cycle.

In other cases, migrants pay huge fees to people-traffickers for urban or foreign work, only to find themselves in indentured positions that are little better than slavery, working out their 'contracts' until they can get their passports back and return home. What little money they do earn will be sent home. These include Asian construction workers and domestic workers in the Middle East and Europe, who have little protection and may end up working as slaves, in the sex industry or in inhumane conditions in factories, for example in the agricultural or garment industries.

While borders are tightly controlled and legal routes are so difficult to access, millions of migrants are stuck in situations of appalling abuse. Even when vast sums have been paid to people-traffickers and agents, migrants still face physical and sexual danger, simply for having been born in the wrong location. Most are trying to improve their lives, as we all do, by moving from our parents' home, village or country.

Some are migrating to save their lives.

An hour and a half's drive south from the bustling coastal town of Cox's Bazar, Bangladesh, is Kutupalong, the world's largest refugee camp. Over a matter of weeks in 2017, the forested hillsides here were razed, as one of the poorest districts of one of the poorest nations became home to around 1 million Burmese Rohingya, fleeing genocidal violence across the river in Myanmar.

The vast, sprawling ghetto is a social and environmental calamity. I visited during the dry season, when the untethered soil and sand streams off the hills in the breeze, and a thick layer of dust coats everything. After just two hours in the camp, my throat was burning. It is worse in the wet season, I was told, when the denuded hillsides become a mud slick and camp inmates are forced to wade through filthy flood water.

Kutupalong is a city constructed of polythene and bamboo shelters interspersed with alleyways running with raw sewage. It is peopled by desperate souls – bereaved, injured and hopeless families. Each person I spoke to had their own horrific tale of trauma and loss, yet the over-riding anguish I heard from everyone was that they weren't allowed to work. Men, women and children were essentially prisoners, whiling away long hours of unemployed boredom sitting on the ground by their shelters. Violence, especially against women and girls, was high, and people-smuggling was a constant danger. There was a thriving black-market economy in the camp, but it neither benefited the wider Bangladeshi community nor the vulnerable inmates, who were instead both exploited and a burden.

Kutupalong has the population and the houses, streets, religious and social buildings of a city; nevertheless it does not function as a city for one overriding reason: it isn't networked beyond the camp boundaries. Cities work as economic hubs because they are concentrated nodes of connectivity – people exchange money and resources, trade labour and combine ideas to create something bigger than the sum of their parts. Restricting any part of this exchange limits the economy, as the UK discovered after its decision to end freedom of movement between Britain and the EU resulted in labour shortages that reduced the availability of everything from food to fuel. Prime Minister Boris Johnson, comparing immigrants to heroin, complained businesses had been able to 'mainline low-wage, low-cost immigration for a very long time' and needed to wean themselves off, but nevertheless was forced to issue emergency visas in a desperate attempt to attract delivery drivers, fruit-pickers and other essential labour. In 2021, job market data from the UK's Recruitment and Employment Confederation found the nation was short of 2 million workers. Brexit Britain became the only example of a Western nation deliberately imposing trade barriers on itself, with a predictable economic outcome.

Although the Bangladeshi government accommodated the vast numbers of Rohingya, it has not granted them refugee status. Without this, they are not permitted to leave the camp or work, and they have limited access to education. As citizens of nowhere, the Rohingya are trapped on a bare hillside in a foreign country with no hope. As if to underscore the division between natives and migrants, the government is now starting to move these vulnerable, stateless people to an isolated island, prone to cyclones and flooding, in the Bay of Bengal.

'We want citizenship,' I was told repeatedly by Kutupalong's inmates. Asylum seekers globally can spend years in this limbo, forbidden to join the formal economy, take part in society or move on with their lives. It's a waste for them, and it turns a national opportunity into a burden. In the UK, for instance, there are around 400 asylum seekers who have been waiting more than a decade for their claims to be processed.

It's even worse elsewhere. I've visited people in refugee camps in different countries across four continents, where millions of people live in limbo, sometimes for generations. Around the world, whether the refugee camps were filled with Sudanese, Tibetans, Palestinians, Syrians, El Salvadorians or Iraqis, the issue was the same: people want dignity. And that means being able to provide for their families – being allowed to work, to move around, and to make a life for themselves in safety. Currently, too many nations make this wish – though it is very simple and mutually beneficial – impossible for those most in need of it. As our environment changes, millions more risk ending up in these nowhere places. Globally, this system of sealed borders and hostile migration policy is dysfunctional. It doesn't work for anyone's benefit.

We are witnessing the highest levels of human displacement on record, and it will only increase. In 2022, refugees around the world exceeded 100 million, more than doubling since 2010, and half were children. Registered refugees represent only a fraction of those forced to leave their homes due to war or disaster. In addition to these, 350 million people are undocumented worldwide, an astonishing 22 million in the US alone, the UNHCR estimates. These include informal workers and those who move along ancient routes crossing national borders – these are the people who increasingly find themselves without legal recognition, living on the margins, unable to benefit from social support systems. Right now, one-quarter of the world's children, half of them African, are invisible, meaning that there is no formal record of their existence.

Sometimes conflict, persecution and natural disasters force people to flee. Sometimes it is a culmination of small indignities – of poverty, joblessness, prejudice or harassment. It is not always clear when a person is down on their luck whether they are a 'genuine' refugee and 'deserving' of asylum – whether they are a 'good refugee' – and such judgemental terms help no one and diminish us all.

As long as 4.2 billion people live in poverty and the income gap between the global north and south continues to grow, people will have to move – and those living in climate-impacted regions will be disproportionately affected. Nations have an obligation to offer asylum to refugees, but under the legal definition of the refugee, written in the 1951 Refugee Convention, this does not include those who have to leave their home because of climate change. Things are beginning to shift, though. In a landmark judgment, in 2020, the UN Human Rights Committee ruled that climate refugees cannot be sent home,[10] meaning that a state would be in breach of its human rights obligations if it returns someone to a country where – due to the climate crisis – their life is in danger.[11] Today, the 50 million climate-displaced people already outnumber those fleeing political persecution.

The distinction between refugees and migrants is not always a straightforward one. Someone fleeing war – or indeed, drought – in their home country, may arrive in another as an asylum seeker and eventually be granted refugee status. But if they move to find work or to unite with their family, they may become an 'economic migrant', a class of migrant that has been politically cast as being less 'deserving' of welcome, and potentially a burden on the social system. While the dramatic devastation of a hurricane erasing whole villages can make refugees of people overnight, the impacts of climate change on people's lives are usually gradual – another poor harvest or another season of unbearable heat, which becomes the final straw that pushes people to seek better locations. Such people would be categorized as economic migrants, but they are also refugees from the Holocene world, the pre-Anthropocene landscapes that their ancestors' cultures and societies learned how to make into a home. With those Holocene environments gone, we are all finding our feet in the Anthropocene, and no one has better claim than anyone else to the habitable twenty-first-century landscapes.

5

Wealth of Migrants

Our recently invented national border controls, and our newly inherited attitudes towards national identity and migrants, are causing us all huge and unnecessary trouble. People are trapped in terrible, often deadly, circumstances around the world; instead they could be helping to improve their lives and the societies and economies of safer locations.

Immigrants expand economies, innovation and wealth. Migrants contribute far more than they take out of societies – part of this is synergistic because they move to cities. And, of course, there are also huge benefits for the migrants: migration is by far the most effective route out of poverty. According to Bryan Caplan, professor of economics at George Mason University, opening borders would lead to 'the rapid elimination of absolute poverty on earth',[1] because people would be able to move to locations with earning opportunities.

Migrants also improve livelihoods, housing, education and opportunities in the countries they physically leave behind because they remain economically linked to them. The networks that migrants forge help with technology transfer, trade, investment and also the transfer of institutions and norms that foster growth. One study found that an Indian entrepreneur citing a patent filed in the US was more likely to cite one filed by an ethnic Indian engineer, for instance.[2] And the same was found for Chinese patent seekers. Knowledge flow is not random, it channels through the networks that migrants make and maintain. Skilled migration shapes the comparative advantage of nations, and boosts the origin nation's economy, directing investment into training and education.

The Philippines, for example, has developed a formidable reputation for its expertly trained specialist nurses, for which global demand outstrips supply. Ageing populations in Western nations have a shortage of dementia-care nurses, and emigrating offers Filipinos the chance to escape poverty. Highly trained people often struggle to find employment in poor countries, and so benefit from migrating to where they are needed, and this option can actually spur enrolment in high school and further education, raising education levels across the entire community. One study found that an increase in migration caused a 3.5 per cent increase in secondary school enrolment in migrants' home communities in the Philippines, and also a general rise in income in the source area.[3] However, origin countries, smaller ones particularly, can suffer from an exodus of highly skilled people, such as doctors. Over the longer term, this 'brain drain' is the exception rather than the rule, according to a large body of evidence,[4] and emigration actually benefits origin countries. Where there are fears of a brain drain, destination and origin countries should work together to tailor skills to benefit both. This means, in the Philippines, ensuring that host countries invest in the training for extra nurses, and in the different specialisms required by host and also origin nations, such as paediatric care, where there is a local shortage. Dementia cases are set to quadruple globally by 2050, so investing in training in the Philippines would enable the nation to become a centre of expertise for its own needs too. Bilateral agreements between countries could ensure that skills and technology investment by labour-poor host countries into labour-rich origin countries meet both nations' needs, as well as migrants themselves – an investment to train 50 per cent of nurses in a college on the pledge that 20 per cent would be offered employment in the host nation, for instance. These agreements could include other social and infrastructure investments in origin countries, too.

In 2018, the Global Skill Partnerships model was included as a policy agreed by 163 nations that signed up to the Global Compact for Migration.[5] In this model, the country of origin agrees to train people in skills specifically and immediately needed in both the country of origin and the destination. Some of those trainees choose to stay and increase human capital in the country of origin; others migrate to the country of destination. The country of destination provides technology and finance for the training, and receives migrants with the skills to contribute to the maximum extent and integrate quickly.

Where nations have experienced a large amount of skilled emigration, they can also take steps to tempt these people back. China and India have both attracted their emigrant talent back through investment in research and development, policy changes, new support of institutions and family perks. Migrants increase their skills and experience while living abroad, and bring back valuable expertise. The acceleration of global knowledge transfer that comes with global migration is an important benefit, and will be a vital part of our transition to the new green economy and to the reduction of poverty.

What this also means is that by putting up barriers to immigration, governments in rich countries are not only forcibly preventing the world's poorest people from helping themselves, they are also hampering their own national productivity. As an overwhelming wealth of research shows, a strategy of accommodating migrants is a far better ploy than attempting to keep them out. The former preserves national (and regional) stability and strengthens a state's economy; the latter leads only to conflict and misery, with ramifications that can last for generations. By contrast, 'better economic and social integration of migrants could lay the groundwork for economic gains of up to $1 trillion globally', according to a landmark report covering more than 200 countries by the McKinsey Global Institute.[6]

What if we thought of the planet again as a global commonwealth of humanity, as it was for millennia, before formalized border controls started to be introduced in parts of Europe a few centuries ago? Enabling global free trade in labour is, according to Caplan, 'not only just, but the most promising shortcut to global prosperity'.[7] Michael Clemens at the Center for Global Development in Washington, DC, estimates that opening the world's borders even to temporary migrant workers would double global GDP. In addition, we would see an increase in cultural diversity, which we know improves innovation, just when it's most needed to solve unprecedented environmental and social challenges.

Removing borders would improve humanity's resilience to the stresses and shocks of global climate change. However, it would also result in losers among some sectors of society, particularly in the host nations, so strong dynamic social policy with a welfare contingency would be required to help the transition.

One of the biggest fears surrounding mass immigration is the idea that immigrants take jobs from the native-born and drive down wages. This sounds plausible but is actually a fallacy, because the economy isn't a zero-sum game. For a start, migrants bring a greater diversity of skills to the workforce, which improves the efficiency of the overall economy, which leads to more jobs. Migrants also increase the size of the economy: they need to eat, shop, get their hair cut and so on – by spending their wages and paying taxes, migrants support new jobs and businesses.

The economic benefits of immigration are significant, immediate and remarkably long lasting. One study found that US counties that received larger numbers of immigrants between 1860 and 1920 had a 57 per cent average increase by 1930 in manufacturing output per capita and up to 58 per cent increase in agricultural farm values; and also, 20 per cent higher average incomes and education attainment, and lower unemployment and poverty rates *in 2000*.[8]

Nevertheless, considerable fears remain, particularly around the influx of low-skilled workers, since these jobs are the ones available to the widest population of native-born people, particularly those who are poorest. The terms 'low-skilled' and 'high-skilled' are used by policymakers and economists to describe the level of formal education that a person's profession requires, with the highest skilled jobs – a cardiac surgeon, for instance – requiring multiple university degrees, whereas low-skilled jobs are associated with manual labour. In reality, the terms don't fully reflect a person's skillset, and ignore many other qualities that are important in employment, such as willingness to work or ability to learn. They also do not necessarily reflect a job's value. However, migrant (and native) workers in these two categories are treated very differently.

Numerous studies have looked at the effect of immigration, and the evidence shows that even large waves of low-skilled migrants arriving have no negative impact on the wages or employment prospects of the native population – and often, they have a positive impact. If you were to compare the wages of natives in cities with little immigration and those with a large share of immigrants, you'd find that wages are much higher where there are *more* migrants. Partly, this is because migrants tend to go where there are better options, so cities that are doing well will attract more migrants. But it is also the effect of the migrants themselves on the economy.

In April 1980, Fidel Castro unexpectedly gave a speech in which he announced new permission for Cubans to escape the country's previously sealed borders. By September at least 125,000 of them, most with very little education, had arrived in Miami, swelling the labour market by at least 7 per cent. David Card, a labour economist at the University of California, Berkeley, looked at the effect of this influx on wages in Miami, comparing them before the big migration, during, and for some years afterwards – and he compared this to wage trajectories in other similar-sized cities in the country (Atlanta, Houston, Los Angeles and Tampa). The migrants hadn't selected Miami for any reason other than that it was the closest landing point to Cuba, and Castro's announcement had been sudden, so Miami's workers and businesses had no time to react.

Card's study found that the wages of natives – even the lowest-skilled workers – were not adversely affected by the rush of immigrants. Since, countless other studies from around the world have concluded the same: there's no evidence that immigrants take jobs from natives or lower wages. These include studies looking at the immigration of 'repatriated' Algerians to France after postcolonial independence in 1962, massive immigration from the Soviet Union in the 1990s after its borders were relaxed, which increased the population of Israel by 12 per cent in just four years, and another study looking at a huge influx of immigrants from across the globe to Denmark between 1994 and 1998. Indeed, one study, by Giovanni Peri of the University of California, Davis, concluded that immigration to the US between 1990 and 2007 had actually boosted the average wage by $5,100 – a quarter of the total wage increase over that period.[9]

Despite continuing evidence from these and more recent studies, some people's fears persist about the threat immigration poses to jobs, perhaps because it seems counter-intuitive to conclude that none exists. Usually, after all, the more of something you have, the lower its price – the rule of supply and demand. However, this doesn't apply to jobs and wages for a few reasons.[10] First, the increase in the supply of workers as migrants join natives is offset by the increase in demand for workers, as the migrants spend money on goods and services that boost businesses. We can see this compensatory effect best by looking at the cases where it is absent. For a brief period, Czechs were allowed to work – but not live – across the border in German towns, where they made up as much as 10 per cent of the workforce. Although these commuters had little effect on wages in these towns, they had a big effect on jobs – there was a large drop in native employment. This was because the Czech workers weren't spending their wages in the host country, they were taking all the money home with them – because they *weren't* immigrants.

Another reason that low-skilled migration actually increases jobs is that it slows down the adoption of mechanization and automation, both of which require huge investments of capital and training, and often alter supply chains. A ready supply of affordable workers, especially for farm and factory work, makes labour-saving technologies less attractive to industry bosses. Whereas when immigrants are expelled or prevented from entering the country, businesses that relied heavily on them switch to mechanization and specialize in only those types of production. For instance, when Mexican farm workers were kicked out of California in 1964, tomato harvesting went from entirely hand-picked to entirely mechanized in two years. Over the same period, California stopped producing crops for which mechanization wasn't available, including lettuce, asparagus and strawberries. In other words, the jobs available to all workers reduced significantly once immigrants left.

Immigration also triggers reorganization of the labour market, which is almost always beneficial to natives. Generally, low-skilled immigrants get the manual jobs, and native workers with local language skills and more experience are upgraded to non-manual jobs that require better communication skills, with higher wages. This kind of occupational upgrading was seen in the Danish studies, and also during the great European migration to the United States at the turn of the twentieth century. In other words, migrants and natives don't compete directly for jobs, but the diversity of abilities, skills and knowledge increases general productivity in the workforce to everyone's benefit. Generally, an increase in migration means that workers can be more efficiently matched to demand and skillset, boosting productivity across the broader economy. Increased labour brings an increase in profits that can be invested in more production.

Importantly, most jobs available to immigrants are the jobs that natives don't want to do, and that oils the wheels of the wider economy. And if immigrants are paid to do jobs of childcare, looking after the sick and elderly, cleaning and cooking, that enables natives (who were formerly doing this as unpaid labour) to also join or rejoin the workforce. In the patriarchal structure of most nations, this has meant that highly skilled women have been able to join the workforce when there were more migrants around.

Enterprising migrants start businesses that employ more migrants and natives. Some of these businesses help generate greater social, economic and cultural productivity, driving a whole new economy. Think of the Chinatowns, Little Greeces and Little Italys, for example. In the past, the vast majority of migrants were the poorest, most unskilled people, who had the least to lose by leaving destitution for the price of passage elsewhere. Today, with border controls and immigration controls, only the wealthiest can afford the cost and only the most skilled and most motivated can legally enter many countries from the poor world. Many bring exceptional talents, ambition and knowledge. They, or their children, become the job creators, innovators and entrepreneurs that drive their host nation's economy. Many are household names. Immigrant-founded companies account for more than half of the top twenty-five American firms and many of the most recognizable brands, including Google (and its parent company Alphabet), Yahoo!, Kraft Foods and Tesla. Silicon Valley is the Geneva of technology: half of the companies' founders are immigrants, as are two-thirds of the workforce. Henry Ford was the son of immigrants, Steve Jobs's father was from Syria, and the Pfizer vaccine against Covid-19 was made by Turkish immigrants to Germany and Hungarian immigrants to the US.

The truth is, we need a diversity of skills and talents for productive economies and vibrant societies, which means everyone from fruitpickers to computer programmers, from truck drivers to ballet dancers.

THE DEPOPULATION CRISIS

In 1950, women were having an average of 4.7 children in their life-time. The fertility rate near-halved to 2.4 children per woman by 2020. But that masks huge variation between nations. The fertility rate in Niger, West Africa, is 7.1, but in the Mediterranean island of Cyprus women are having one child, on average. In Europe, the aver-age is 1.7, which presents a huge challenge.

Europe's population is set to shrink by 10 per cent by 2050. It is also ageing. The elderly and children will outnumber workers by one-fifth by 2060. Germany alone would need to bring in at least 500,000 migrants each year just to offset that. Population growth in the UK, too, is entirely down to international immigration, according to the Office of National Statistics – the national fertility rate is just 1.7.[11] The rosiest forecasts for the next decades see Europe's working-age population falling by one-third. Already, almost two dozen countries are getting smaller every year, from Poland to Cuba to Japan, which lost almost 450,000 people in 2018. In these countries, women have fewer than the averaged 2.1 babies that would allow for a population to remain stable. The population decline would be even steeper were it not for steadily increasing life expectancy, but it will catch up. Japan is expected to go from 128 million people to fewer than 53 million by 2100.

This will profoundly change global society. The post-war baby boom – boosted in some countries by the post-war 'economic miracle' – created a world dominated by young people with political and economic power in the 1960s. It produced a decade of huge social change and disruption. The population swell of these baby boomers is still making its way down the generations, but it's clear we are now entering an ageing world, where elderly people hold the wealth and power – in the UK, one in five baby boomers is a millionaire.[12] By 2065, more than one-quarter of the population will be over sixty-five years old. Grey societies generally experience less violence and war, but they are also less collaborative and more protectionist. It was the grey voters that ushered in Brexit, Trump and Erdoğan.

This demographic problem is a significant crisis affecting most of the world's nations: without the productivity of young workers and their taxes to support increasing numbers of long-lived elders, as well as children and those who cannot work because of sickness or disability, society cannot function and the economy would stagnate or crash. China is expected to reach a turning point around 2025, with its population plateauing or even shrinking as fewer babies mean a slowdown and then a reversal in the momentum of economic growth. Its fertility rate has slipped to 1.3 children per woman, and families are already importing brides – sometimes by kidnap – from poor villages in neighbouring nations to be carers for the groom's ageing parents. Declines in birth rates in other large countries have also been extreme: India is at the replacement rate of 2.1 and falling; and Brazil, the fifth most populous country, has a fertility rate of 1.8. Russia's depopulation crisis has emptied rural towns of their youth, while across Europe entire villages have been put up for sale or touted for free in the hope that someone will revive them. Shrinking cities get less investment, which speeds their decline and makes them less attractive to new residents – a downward spiral. In the US, cities experiencing sharp negative growth include New York, San Jose and Boston. In the EU, 20 per cent of cities are shrinking, posing a huge challenge.[13]

The upshot is that demographic shrinkage means the US will need at least 35 million more workers by 2030; and by 2050, the EU will need 80 million extra workers, and Japan will need another 17 million to be able to maintain existing living standards and social support systems. Even during economic downturns, these economies have experienced substantial labour shortages, eased somewhat through amnesties of 'illegal' migrants. However, states will soon have to compete for low-skilled migrants, let alone skilled ones. It's worth noting that, despite perceptions, the bulk of the immigrants entering through points-based entrance systems (in which skills are awarded points and migrants gain entry if they have enough of them), which are operated by countries like Australia, are actually low-skilled because they enter as family members of the desired 'high-skilled' migrant. Most migration to Europe is driven by women working in domestic service.

Populations are continuing to grow in some parts of the poor world, particularly on the African continent, although in many countries these rates are also coming down. Nevertheless, Africa is heading for a triple whammy of exploding numbers of young people living in rural poverty at the mercy of environmental catastrophe. South Asia, where nearly one-quarter of the global population lives, faces similar issues, and will soon have the highest prevalence of food insecurity in the world, according to the World Bank. Some 8.5 million people have fled already, mostly to the Persian Gulf, and as many as 36 million more may soon follow, according to the Bank. Many are expected to settle in India's Ganges Valley, but this can only be temporary because by the end of the century heatwaves and humidity will make that area uninhabitable too.

The solution is so obvious it hardly needs spelling out, and yet it is rarely discussed as a serious policy: help people move for everyone's benefit.

Many of the countries affected by climate stress and other pressures, such as repressive regimes, have large numbers of unemployed youth living in poverty, which triggers conflict. Creating secure migratory pathways to depopulating, safe countries would help these people, who are often highly educated, get on with productive lives. During the Syrian crisis of 2015–16, Germany and Sweden took in large numbers of refugees, and benefited. The majority have been able to find jobs in their new homes, despite only 1 per cent arriving speaking the language, and although there was a rise in ethnic tensions, elections in 2021 showed a marked *decrease* in the popularity of anti-immigrant far right parties.

For Germany, the acceptance of 1 million refugees, while a generous response to the humanitarian crisis, was an astute economic decision: the country needed to fill labour shortages, partly resulting from the depletion of its Turkish migrant population, many of whom had returned to their homeland during its economic boom. Sweden also took the opportunity to revive its depopulated villages, including reopening schools and football teams. The biggest fear Sweden faces is those immigrants leaving and returning to Syria. With an ageing population there's an economic imperative to increase immigration – to keep the elderly dependency ratios low. I have no doubt that those Ukrainian refugees who end up settling in their host nations will also prove economically beneficial to these countries in the long term.

Migration also benefits the migrants, of course. On average, a migrant moving from a poor to a rich country can earn three to six times what they could at home, according to the World Bank. An unskilled Nigerian worker can earn 1,000 per cent more in the US than in Nigeria; a Mexican labourer earns an average 150 per cent more. Better pay is one of the drivers of migration, but the opportunities afforded through mobility are more than just financial. Rich countries often have superior institutions, better and less corrupt governance, more efficient markets and well-run global companies, and they are safer. This environment means that workers are more productive in a rich country than a poor country, even doing the same job. Scientists are more productive in rich countries because they have better equipment, more stable funding, a wider selection of expertise to draw from, more opportunities to collaborate, and so on. Construction workers in rich countries build better buildings because they have access to better tools and higher quality materials, reliable supplies of electricity and water, and more stringent, enforced regulations governing safety and quality of the build.

TRILLION DOLLAR BILLS LYING ON THE SIDEWALK

If rich countries increased their population by just 3 per cent through immigration, they would boost global GDP by over $356 billion in less than a decade, according to a World Bank study. The lead researcher, Michael Clemens, said that there were effectively 'trillion dollar bills lying on the sidewalk' to be picked up if borders were all opened.[14] It makes sense: to grow an economy, you need to increase productivity, and one way to do that is to increase the labour force.

A study by the UN's International Labour Organization of fifteen European countries found that for every 1 per cent increase in a country's population through migration, its GDP grew by 1.25–1.5 per cent.[15] Australia managed to grow its GDP by 3 per cent during the 2009 global recession partly because it imported migrants – one in four Australians was born overseas.[16]

So migration isn't just an adaptive strategy for individuals, but also for societies. One study estimated that many of the economic costs of climate change could be mitigated simply by moving economic activity in response.[17] Currently, 90 per cent of global production uses just 10 per cent of global land, so moving the climate-endangered 10 per cent to more hospitable environments within the 90 per cent makes sense and is very doable, according to the study. Researchers modelled the global economy under different temperature increases, and under one scenario allowed people to move around the globe freely, and in another restricted their movements. Welfare losses were small in the first scenario, they found, because there were big movements of people northwards: a relatively small increase in temperature of 2°C at the equator (so 6°C at the North Pole), results in a shift of ten degrees north in latitude for the average locations of agriculture and manufacturing by the end of the century. That's like Oslo having the climate of Frankfurt, or Chicago having the climate of Dallas. A greater rise in temperature results in a bigger locational shift.

However, in the second scenario, where movements were restricted, welfare costs dramatically increased. When the modellers introduced a hard border at the 45th parallel (passing through the northern US and southern Europe) with around 1 billion living above and around 6 billion below, those above the border experienced an increase in agricultural productivity, while those below soon became 5 per cent poorer. In other words, facilitating migration can make our economies more resilient.

Rural living is the single largest killer of humans today, owing to poorer access to healthcare, clean water and sanitation; greater levels of poverty and malnutrition; and riskier livelihoods. Three-quarters of the world's hungry people live in rural areas, according to the International Fund for Agricultural Development (IFAD), and, on average, wages are 1.5 times lower in rural than urban areas. The problem is treatable through urban migration.

Today, hundreds of millions of people are migrating from their ancestral villages to rapidly swelling cities in the biggest migration in human history. This means leaving strong intergenerational networks and often family-owned land on which some food can be grown – but where, usually, there is little or no wage or opportunity for development, and deepening poverty. With each generation, family land is split into smaller and smaller parcels, from which food must be grown during worsening environmental conditions. This alone is enough to prompt the desperate to seek their fortunes elsewhere. By 2100 we will be an almost entirely urban species.

The son or daughter who heads to the city for work, doesn't usually migrate permanently – not initially, anyway. They move to earn money to send to their family back home, usually supporting dependants, including ageing parents, their own children who will be looked after by relatives, and often younger siblings. The remittances that these urban migrants send home are what enable the agricultural economies of many nations to survive. Annual global remittances are around $550 billion, although they decreased during the pandemic. Bluntly, if it weren't for remittances from migrants, many villages in the global south would be unsustainable and deserted already.

These remittances have broad benefits that extend well beyond the migrants' immediate families. In Ghana, children whose families get help from a relative abroad are 54 per cent more likely to attend secondary school. The schools themselves are often built or supplied with the help of remittances, and money sent back home enables people to build houses, start businesses that employ other people, and invest in machinery or upgrade equipment.

Key to this is the circular network that enables migration and the prosperity it fosters in host and origin community – this circular, back and forth movement of people and their resources is a hugely important part of sustainable migration and needs accommodating in any successful policy.[18]

The people who move create more diverse cities of opportunity where they arrive, easing the transition for others, and meanwhile improve livelihoods, housing, education and opportunities in the villages they leave behind.

Migrants who go abroad to university encourage those back home to go to school, studies show, and those who end up moving to democratic nations help promote democracy back home – by encouraging friends and family members to vote, for instance. One study, in Mali, found that returning migrants were far more likely to vote, and also that voter turnout rose for non-migrants in neighbourhoods that had large numbers of returnees.

Indeed, migration is by far the best and most efficient way to help nations achieve pretty much any indicator of development, and makes much better sense than most aid spending, even if politically it's usually less palatable. For most developing countries, the amount of money they receive in remittances outstrips what they receive in aid from rich nations by 2.5 times on average. Nigeria, for instance, got $24.3 billion in 2018 from its emigrants working abroad, which is eight times what it got in development aid and more than ten times what it received in foreign investment. Aid budgets get carved up into administration, salaries, campaigns and buying four-wheel-drive vehicles for aid workers. Remittances, however, are sent directly to recipients and invested in ways that improve people's lives, even if money agents and currency transfer fees swallow up a disproportionate amount. UNESCO calculates that lowering these fees could boost private spending on education by $1 billion a year.

Nevertheless, among the more humane ideas that rich nations alighted on in their attempts to deter migration was to tackle the 'root causes'. To that end, the EU, for instance, set up a multi-billion-euro fund for aid for Africa. However, several studies find that neither aid nor economic development in poor countries reduce migration – in fact, sometimes they do the opposite. Migration actually tracks well with development: as poor countries get richer, their rate of emigration rises. This is in large part because migration to a rich country is expensive – it costs a lot for a people-smuggler, plane tickets and other routes out of the country, such as higher education. The very poorest can't afford it, and the ones who can see it as an investment in their future. As countries become wealthier, more people are able to afford that investment, and this continues until the average annual income per head reaches around $10,000, at which point the small rise in comparative earnings with rich countries makes the costs and upheaval of migration less desirable. Currently, the average income in sub-Saharan Africa is one-third of this.

None of this is to say that development aid should be abandoned – it is vital to helping improve health and education in poor countries, which is an obligation of rich nations, particularly given their historical role in impoverishing these countries through colonial practices, resource extraction and other policies. However, spending on aid to stop migration is like printing more textbooks to stop people going to school – it's misconceived. One study calculated that to deter a single migrant from a country like Iraq coming to an EU nation would cost $1.8 million in aid, and to deter those coming through official channels would be even costlier, between $4m and $7m per person.[19]

Equally, if we look at global productivity purely from an economic perspective, labour is the most important commodity humans have. Sending money to Nigeria rather than enabling Nigerians to move to where the work is, makes about as much sense as trying to farm the deserts or produce factory cars in Antarctica. Similarly, it is harder and far less efficient to send stable, established institutions, peace, prosperity and good governance from the Netherlands or Canada to Sudan or Yemen, than it is to translocate workers and their families to more productive countries. Migration is not just inevitable but should be encouraged. During the limited decades we live on this earth, people should be free to move to locations with better opportunities and not be trapped purely by accident of birth.

Clearly we're a long way from such flexible borders today, so how might we approach a well-managed system of global mobility? Like climate change, mass migration is an issue that must be managed at the global level. The globalization of human activity and the planetary scale of the issues we face demand a new era of cooperative bodies with teeth to act. We have experienced the consequences of decadal erosion in the powers of global bodies, from our failure to act on greenhouse gas emissions to our failure to vaccinate the global south against Covid in a timely way. This should spur us to strengthen global cooperation over international problems.

Ideally, all migrants would be matched with job vacancies before leaving home, but in reality many will be forced by extreme events to move more urgently. One way to address this problem would be by establishing a global UN Migration Organization with real powers to compel governments to accept refugees (which they are anyway required to do but often don't); to agree a sensible plan for the redistribution of people whose circumstances are becoming harder (owing to climate change) before they reach an impossible crisis; and to manage both the immediate and long-term strategy of relocation, remuneration, funding and, potentially, returns. This body would need to be set up with intergovernmental backing, be run by an international consortium of civil servants advised by experts (including social scientists, town planners and climate modellers), and be fully funded by an international 'taxation' system contributed to by all nations.

One idea would be for nations to agree to migratory quotas, receiving agreed funds or loans for the initial, short-term social and economic integration costs of accepting new people, including the investment in larger cities. Some of this could come from the origin nations. The European Union has been trying to push for some sort of quota system for refugees and asylum seekers for some years, only to be blocked by member states including Hungary and Poland (ironically, the least desirable nations from the migrants' perspective). Consequently, the asylum process in the EU is utterly broken and rather than welcoming these migrants as the social and economically useful contributors they are, they have instead become a social and economic burden endured by a few southern states on whose borders they wash up. The migrants live in limbo, unable to work or build lives in places where they have network connections. As I have described, they are often held in inadequate camp prisons for years; locals become resentful, and people die needlessly.

Instead, everyone could be offered United Nations citizenship in addition to their birth citizenship – for some people, such as those born in refugee camps, lacking papers, or citizens of small island states that will cease to exist later this century, UN citizenship may well be their only access to international recognition and assistance. Passports would be issued on the back of this, perhaps based on the stateless persons' passports, also known as Nansen passports after their Norwegian promoter, the polar explorer Fridtjof Nansen, who was the first international High Commissioner for Refugees. Half a million of these internationally recognized refugee travel documents were issued after the First World War, from 1922 to 1938, the majority to Armenian and Russian refugees. The celebrated photojournalist Robert Capa, originally from Hungary, was one bearer of the Nansen passport.

The Nansen passport was valid for up to one year, although it could be renewed. It enabled holders to travel to another country to look for work, thereby helping relieve the pressure on overcrowded places and distributing refugees more 'equitably' among member countries of what was then the League of Nations, while enabling states to keep track of displaced people entering and leaving their borders. For refugees, it provided a new form of international protection, over and above the authority of the asylum state, before they acquired a new citizenship. There is a lot to admire about the Nansen scheme, particularly today, when refugees and other migrants are trapped, unable to make safe border crossings or find work. A Nansen-style emphasis on enabling mobility, so migrants are free to move to where they can find work, is vital to safe and positively managed migration – additionally helping states with the data they need to manage changing demographic needs – and it's something the UN-citizen scheme could address.

People would apply for migration, ideally in their country of origin, and be offered place-based visas in safe cities, according to the quota system. Migrants with sought-after skills or wealth would inevitably find it easier to choose their destination city, but the quota system should be set to ensure safe homes for everyone in need. Schemes, such as temporary or job-specific visas, or a migration lottery, can be used to help allocate migrants within the quota system. Many countries already use some form of immigration lottery, including the US, UK and Canada.

Given the scale of this migration and the speed at which we must undertake it, migrants will have to be involved in building their own cities or expanding existing ones. Visas could include a requirement that the bearer commits to a certain number of hours a week for, say, two to five years of community service, which, depending on age, skills and ability, may include working in construction, caring jobs, litter and waste management, wildlife restoration, and other necessary occupations that a nation needs to fill. Training and salary would be provided, and migrants could be given ownership options on the resulting homes and business spaces. This would help ease the cultural and social transition of new citizens, building a progressive civil society, especially if native-born people were also committed to such programmes.

Given the difficulties with establishing any sort of globally binding agreement, we shouldn't wait: bilateral or regional arrangements should be rolled out, especially where cultural or historic links already exist. Pacific nations already have agreements with Australia and New Zealand, for instance, and other regional groups have agreements over reciprocal labour rights, recognition of skills and qualifications, and freedom of movement. The best model of this is the European Union with freedom of movement, including trade and labour, across its member states, and in which, if southern Spain suffers an unbearable heatwave, residents are free to move northwards to less affected countries. The African continent is working towards a similar system of free movement across all nations as part of its Agenda 2063 development initiative, which includes an African Union passport and free-trade agreement. The free-trade pact is now at an advanced stage, with fifty-four of fifty-five nations signed up (all except Eritrea); the free movement protocol is progressing more slowly, with thirty-three nations so far signed up, although several nations have begun offering visas on arrival. Analysts expect the free trade and movement revolution to transform the continent's economies, and enable its bulging youth population to find employment – average unemployment fell by 6 per cent in Europe due to free movement within the European Union, for instance.

It's 2032, and Ajay Patel is applying to migrate. Originally a rice farmer from rural Gujarat, India, the family migrated first to the city of Ahmedabad, then to Mumbai, after drought and rising sea levels increased the salinity of the soil, making agriculture impossible. He and his wife have three teenage children and live in a slum shelter, where they are street hawkers. The slum is regularly flooded during the now frequent violent storms, and regular heatwaves are making conditions deadly – back in 2020, the slum was 6°C hotter than the rest of the city; these days, the heat combined with post-rains humidity are proving fatal. Ajay makes his family application to the UN Migratory Office in Mumbai, listing his details, including the family's skills, and up to three city choices: he lists Manchester (where he has a distant family connection), Glasgow (where he has a friend) and Ottawa (where there is an affordable business school for his oldest son). In addition to their Indian citizenship, they are each issued with a UN passport, which permits them to enter any nation, and to work and live in a large number of them, although not necessarily to receive social security support. For that, they must wait for allocation.

Within a few months, they have been allocated five-year migration visas for Aberdeen. They can accept or appeal against the decision, or refuse it and try their luck elsewhere. They decide to accept. The visas come with conditions: Ajay and his wife must work for at least two years in one of the government's specified sectors, which may include undertaking initial training. These jobs are open to any UN-passport holder, but they must be preferentially given to citizens and holders of migration visas. The children must attend education or other training. They must stay in the country for at least twenty months of the first two years. In exchange, the family gets transport to Aberdeen, housing, healthcare, language lessons and other assistance. After two years, they can choose any form of occupation – Ajay wants to work in retail, eventually opening his own shop. At the end of five years, they can apply for citizenship and have the same rights as any native-born. Those who possess UN passports can also live and work in Aberdeen, but they are not eligible for preferential job allocation or free public services unless they apply successfully. Cities must also accept and support refugees who are unable to work on compassionate grounds, according to the quota system.

Ajay finds work doing energy-efficiency retrofits for buildings, and his wife works as an assistant in social care. Both receive free language lessons and within a couple of years Ajay is helping train new immigrants, and his wife is enrolled in a part-time course to become a carer. Ajay is hoping to open a store supplying insulation with a couple of his colleagues. The children are in education.

This is just one hypothetical scenario with one model of rules and restrictions that would enable migrants to move away from danger and poverty, to get on with their lives while contributing to their new society – and provide an adjustment period to win over a sceptical host community. It enables cities in climate-hit parts of the world to concentrate their funds on supporting a smaller population to live in what are otherwise unliveable conditions, through air-conditioned, flood and storm-proofed buildings, for instance. Mumbai cannot safely house and feed its population of more than 20 million people under the climate conditions we face in coming decades – much of the population will need to migrate. The scenario also enables host cities to better integrate much larger populations as they grow, while meeting the labour needs of the vast programmes of climate adaptation and social and infrastructure improvements required over the coming decades.

There are many alternatives to this model, and if we are to manage the certain migration of large populations over the next decades, we have to proactively plan ways of doing so that maintain people's dignity while addressing the huge challenges we face in adapting to a hotter, more hostile world.

6

New Cosmopolitans

People are moving, ready or not. We can and must prepare.

There's no humane way to keep desperate migrants out of the United States or Europe or Australia. Those of us living in parts of northern Europe, Canada and other climatically 'luckier', still-habitable zones will not be able to keep migrants out with walls or guns. They will be too many and they will keep coming because there will be no other choice. The question is whether they will be helped, or whether the rest of the world will stand by and watch them die.

If this sounds scary, it is because the dominant migration narrative we've heard has been centred on the threat of mobs of foreigners over-whelming us, rather than about the opportunities, practicalities and realities of continual human movement. We need to change the narra-tive, we need to recognize that we are all part of this story – we move for work, pleasure, better opportunities for our children and, if we're unlucky, to escape danger. People are also coming here, ready or not.

In November 2021, one of several overloaded dinghies carrying migrants from France to England sank crossing the dangerous English Channel, and twenty-seven people died in the frigid water, including three children. Some people responded to this avoidable tragedy with glee. Protesting fishermen blockaded a lifeboat station, preventing the crew from embarking on a rescue mission to another migrant vessel in trouble. The UK government, which had removed virtually all other routes for people to seek asylum in Britain, talked about deploy-ing the military to send the small boats back to France. Online news articles about the deaths carried hateful comments, and when com-menting was switched off by the publication, the articles were reposted on Facebook with laughing emoji symbols.

A journalist, Ed McConnell, decided to track down some of the people who had reacted to the news of drownings by posting happy emojis, to ask them why.[1] Among the hateful bile McConnell encountered, were claims that the migrants were 'rapists, murderers and terrorists', 'robbing our country', 'taking jobs' and 'straining the National Health Service'. There was also denial and ignorance about the legal rights of an asylum seeker to claim asylum in whichever country they choose.

Decades of anti-migration rhetoric and misinformation means there is massive misconception in rich nations about the basic facts of migration. In Italy, surveys show that on average people think that 26 per cent of the population are immigrants, whereas it's actually 10 per cent. People also believe racist and prejudicial tropes about migrants, particularly when they are repeated by prominent politicians. These include that migrants are often criminals, violent and dangerous. Surveys in Western nations show that it is widely believed that immigrants are poorer, less-educated, more likely to be unemployed and to live on benefits, and to be Muslim and male, than they actually are. In fact, half of all migrants worldwide are Christians, despite Christians making up just one-third of the world's population – they comprise three-quarters of the foreign-born people living in the US and 56 per cent of those living in the EU. Only 27 per cent of migrants worldwide are Muslim, who most frequently move to Saudi Arabia or Russia.

The idea that immigrants move to rich countries just to take their benefits also fails to stand up. The vast majority of migrants move in search of work, and tend to go where the jobs are, not the most generous benefits. Bear in mind that more than one-third of migrants move between developing countries that offer little or no social benefits, simply following job opportunities. Migrants to rich countries are less likely to be in receipt of benefits than natives, partly because they tend to be younger, healthier and motivated to stay in the country for work, returning to their country of origin before they are old enough to need social security payments. In many rich countries, migration controls mean they are prevented from even applying for benefits, and so 'illegal' immigrants often end up paying taxes but not seeking benefits for fear of discovery. In the US, social security paid by employers on behalf of migrants but never claimed by them swelled the US coffers by at least $20 billion over the 1990s alone. Meanwhile, Trump's

2020 restrictions on work visas cost the US economy $100 billion.[2] Overall, the Organisation for Economic Cooperation and Development calculates that immigrants pay at least as much in tax as they receive in benefits. Indeed, the UK's Office for Budget Responsibility calculated that if the country accepted twice as many migrants, it would significantly slash the national debt.

Fears around crime and violence are similarly unfounded. Studies show no increase in crime linked to migratory patterns, except in a handful of cases in which petty crimes rose slightly, where there was a wave of asylum seekers in the UK who were not allowed to work. By contrast, immigrants in the US are far less likely to commit crimes than those born in the US, with one study suggesting that a rise in immigration in the 1990s may actually have driven the overall drop in crime rates that occurred over that time.[3]

Today, from Europe to Asia to the US, we are experiencing a period of potent hostility to immigrants. Over the past decade, promises to reduce immigration and strengthen borders have been a vote winner even for liberal governments in progressive democracies. Meanwhile, populist, nationalist leaders have become ever more draconian in their dealings with foreign workers and refugees. From the mass incarceration of Mexicans in the US, to the hostile environment of Brexit Britain, to the weaponization of Middle Eastern refugees by Belarus in the winter of 2021, the issue of migration has been shaped into a public threat – a crisis – and this has given rise to anti-immigration movements and far-right political parties.

Much of this has happened at a time when the numbers of asylum seekers entering Europe actually declined. (Europe faced fewer asylum seekers in the five years of 2011 to 2015 – even taking into account the extraordinary number of Syrian refugees – than in the last five years of the twentieth century.) The number of people seeking asylum in the EU each year fluctuates, but the bloc usually receives applications numbering in the hundreds of thousands.[4] So the EU, which has a population of 445 million, could hardly be described as under siege. In 2022 the EU actually did experience a crisis on its doorstep, as war forced millions of people to flee Ukraine into neighbouring nations, the closest of which had been the most strongly anti-migrant. Interestingly, these previously hostile nations, such as Poland and Hungary, were generous and welcoming to Ukrainian refugees even though the numbers involved were orders of magnitude greater than the 'migrant crisis' those nations experienced during the few years prior.

There is a strong correlation between migrant levels in a country and positive attitudes towards them, according to a study of twenty European nations. 'Countries with a negligible share of migrants are the most hostile, while countries where migrants' presence in the society is large are the most tolerant,' the researchers found.[5]

The study showed that the migrant crisis is born of factors unrelated to migration, such as questions of institutional trust, social disengagement and political disaffection. The researchers found that people in countries with low levels of institutional trust and social inclusion fear migration the most. Far-right and populist groups regularly link anti-immigration rhetoric to traditionally left-wing economic and social policies, such as job retention and support for the welfare state, which feeds the perception of immigration as responsible for the social problems facing working-class communities. 'Anti-migrant attitudes have little to do with migrants,' the researchers wrote.

Nevertheless, such attitudes are prevalent across our societies and shape policy. This is already problematic when nations are only receiving tens of thousands (or fewer) asylum applications. In the coming decades, many will be receiving applications in at least the hundreds of thousands. Consider that Russia's invasion of Ukraine created 10 million refugees in just the first three weeks. It's to be expected that people will have concerns about mass migration, particularly in places with small, fairly homogeneous populations. These fears are a significant issue, and must be tackled if climate migration at scale is to be peaceful and successful for migrants and host communities.

Managed well, mass migration will be part of life and cosmopolitan societies will be considered unremarkable by the generation who do not remember more homogeneous times. Younger, urban people are already comfortable with a far greater diversity of people than their grandparents are, and this is partly a reflection of demographic change: in the US, just 18 per cent of the post-war baby boomer generation is non-white; whereas almost half of Gen Z, those born 1997 to 2012, are of Black, Latino or Asian heritage. Younger generations are far less likely to view nationality in racial terms, and surveys show that fewer than half of US under-40s think a person's nationality is important, while just 20 per cent think birthplace is important in nationality.[6]

This century is *all* change. The coming decades of environmental change will additionally wreak their own socio-political disturbances, with disruptions to food availability and other significant challenges. So when looking at this future the baseline shouldn't be thought of as your current life as lived today – the comparison rather is between your future city embroiled in climate-adaptive infrastructure changes, a hotter environment with flash floods, more violent storms, poor food availability, a shrunken workforce with little elderly care, a social environment of fear with increased conflict, terrorism, famine and death broadcast to your screens from the global south . . . or far less of the misery, but many more foreign people living in denser cities.

The latter option is far better for everyone. But that doesn't mean it won't be troubling, especially for people who strongly identify with their local town's hitherto homogeneous population, who may worry about a loss of culture when a large number of Asian, African or Latin American migrants move into their town and it becomes a city. Or when their small prosperous town becomes a refuge for large numbers of poor people from southern states. After all, mass movement involves significant change and, naturally, people feel uncomfortable or anxious about this. Managing this transition will be key to its success, and that means addressing concerns before they produce conflict.

So let's explore (and debunk) some of the fears around immigration in more detail.

There are obvious triggers: immigration can put pressure on host communities when local housing, schools, healthcare and other services become strained. This can be avoided through careful planning and adequate investment from governments to manage the costs and delivery of services for the enlarged population. Considering many are failing to provide this for their current citizens, tensions are bound to rise unless it is addressed. The US, for instance, spends just 15 per cent of its budget on social services – half of what the average EU nation spends. This needs to increase everywhere, but particularly in the US, because provision is woefully inadequate and doesn't even cover universal healthcare.

Social change really is difficult. This is because, although diversity increases innovation and leads to more productive outcomes, it requires more investment in mental energy, apart from the other costs. It's far easier to go along with the same ways of thinking and doing things when everyone else thinks and does things the same way as you. Understanding other points of view, thinking about different ideas from a new perspective, can be hard work, even if the rewards are more than worth it. As a result, investments of time and money need to be made in host and immigrant communities to help with this transition. Programmes such as widely available free language classes and a system of mentoring and support for new arrivals, would help.

Mass migration from badly run or failed states raises the widely publicized fear of an increase in crime and terrorism – that immigrants will import their inter-ethnic disputes with them. Stricter immigration and visa policies are a common reaction to terrorist incidents. In fact, the opposite is true: migration is more likely to reduce terrorism than increase it, according to a large study of 145 countries looking at three decades worth of data.[7] The scientists say the main reason for this was because migration increased economic growth. In a few European countries, migrant populations were more likely to commit petty crime than local populations, but this was where the migrants were more likely to be young males; when they were compared to the same local demographic, the migrants were in fact no more likely to get into trouble than their native peers. Immigrants who settle well and are able to work are unlikely to be extremist or terrorists. Native communities whose fears of immigration are not addressed, however, risk producing their own terrorists, including in the form of white supremacists.

Some people, particularly in parts of Europe and northern Asia with relatively small and homogeneous populations, fear the influx of lots of dark-skinned people literally changing the face of their country. It will happen, and it has happened before. As we have seen, the pale skin of Europeans is very recent in human evolutionary history. The original Americans, Europeans and Britons were dark-skinned until their land was colonized by pale-skinned Eurasian steppes people some 5,000 years ago in Europe, and by their descendants more recently from the sixteenth century in the Americas. Population expansion since meant that by 1900 predominantly white-skinned Europe made up a quarter of the world's population, and three times that of Africa. However, by 2050 Europe is predicted to hold just 7 per cent of the world population, and a third that of predominantly dark-skinned Africa. This is the demographic shift we've already discussed: Europeans today are having fewer children than Africans. Cities everywhere are already far more diverse – London's population, for instance, is already 40 per cent darker-skinned. The majority of people in the US will have darker skin by 2040[8] – for many cities and counties, this has already occurred. Multiple studies show that people who live in diverse cities are more accepting of immigrants of all skin colours; whereas those who live in 'whiter' places where they encounter fewer people with darker skin are more hostile to immigrants.

Fear of this transition is driving anti-migrant sentiment through both deliberately prejudicial policy and unconscious bias in society. You may not personally hold prejudicial views against foreigners or people with darker skin, but you will know someone who does and it is very likely you live in a society whose institutions and structures favour lighter skin over darker. This bias is corrosive and needs to be challenged. It is notable that EU asylum policy has never been so generous as when the refugees involved were pale-skinned Europeans who looked and dressed like most EU natives.

Fear and prejudice against people with darker skin is real, cannot be ignored, and must be addressed systemically if mass migration from tropical to northern countries is to be successful. African and Asian migrants who were living and studying in Ukraine when the Russians invaded found themselves treated very differently at the borders when they fled the country. Although EU leaders clarified that all refugees fleeing Ukraine would be given asylum irrespective of nationality, dark-skinned people faced significant difficulty and hardship even leaving the country, let alone in their new host nation. Today, much of the anti-immigration rhetoric plays on (mainly older) people's fears that their 'race' will be overrun by a different 'race', which, as we have seen, is biologically nonsense. If you have pale skin, blue eyes and light-coloured hair, then you are in a minority globally, but this appearance won't disappear. People will still be born with these features; however, your children or your grandchildren may well have darker skin than you. The biggest concern we should have during this transition is ensuring the population maintains a sufficiency of vitamin D through dietary supplements.

It's important to recognize that prejudice is often a defensive reaction based on fear. The very globalized world that gives emigrants and immigrants opportunities has also produced a global elite of cosmopolitans whose passports, financial privilege and education help them take advantage of comfortable mobility – it can seem as if the globalized world has itself migrated on from sedentary people who struggle to make ends meet and feel powerless in a wash of mobilized humanity. Western nations persist with the myth of meritocracy, meaning those whose circumstances allow them to easily migrate between cities believe they have earned this privilege – and with it, their enlightened liberal politics and acceptance of migrants – whereas those whose jobs and lives have stagnated are believed to be undeserving, lazy and backward in behaviour and belief. This itself is prejudice that needs challenging.

Equally, when people struggle or find themselves out of work, it may be easier to blame migrants than deep structural inequalities or themselves. Motivated beliefs help people to rationalize their racism to the extent that when refugee children die seeking safety, this is blamed on the helpless adults who accompanied them, rather than the rich nations that neglected to save them. Migration isn't the problem; poorly designed policy is. The way to solve fear-based prejudice is to address the despair and anger of 'left behind' natives whose wages have declined or who live in communities with high levels of unemployment. That means designing and funding social policies that help people live dignified lives and reduce inequality – and that means using taxation competently as a redistribution tool. For instance, having a high top rate of tax reduces inequality after tax but also *before* tax because it eliminates very high salaries, as they become pointless. Gross inequality isn't an inevitable result of technological progress or capitalism (or any other economic system), but a failure of social policy, including a failure to tax wealth and prevent tax evasion. Smart policy will be needed to target prejudice and promote inclusion during our rapid mass migration.

Inclusion is key. So is it better to house migrants with others from their country of origin, risking creating ghettoes and triggering an exodus by natives (sometimes called 'white flight')? Or to do as in Singapore, where 80 per cent of the population lives in public housing and strict quotas ensure that each building has an ethnic mix? From global studies, the answer is a bit of both: the way to prevent segregation is build public housing for low-income residents and make sure they are dispersed throughout the city so there are no 'pure' rich-only or native-only neighbourhoods, and also to recognize that migrants gain network benefits from a certain degree of social clustering – so enabling people of similar origin to migrate to the same cities can help hugely with social and economic well-being. New arrivals need support to become part of society, and immigrant inclusion programmes can help with this. Bergamo in Italy runs an Integration Academy for asylum seekers – a one-year 'boot camp' with language classes, internships at local factories and businesses, and free community service. It's been criticized for its strict policies, which include uniforms, and for its implicit messaging that migrants should show gratitude for asylum (which is an international human right), but it's also helped migrants to become more employable in their new home, and perhaps as importantly, convince a vehemently anti-migrant public and government that immigrants are a useful and valuable part of society, beyond the humanitarian argument.

Mass immigration brings fears for the potential for radical change beyond what native-born people might feel comfortable with. For instance, anti-immigrant activists often point out that an Islamist government could in theory be democratically elected. However, generally, the more diverse the population is, the less extremist is the government. There are ways policy could prevent fundamentalist or socially regressive rule; suggestions include only extending voting rights to immigrants after a period of a few years, allowing immigrant and host communities to have time to make some cultural adjustments.[9] The children of migrants who have been included in society – the second generation – are generally more politically, sexually and religiously liberal than their parents. Alienation and exclusionary social attitudes, though, can leave second- and third-generation immigrants vulnerable to extremist ideology.

Things will change. The England of the 2020s is not the England of the 1950s and will not be the England of the 2070s. The United States of the nineteenth century is not the United States of the twentieth century or of the twenty-first century. Places change and immigrants play a large role in that. The alternative to cultural expansion and change is not stasis but regression and, in extreme cases, death – as the Palaeo-Eskimos showed us.

Mass migration will be an upheaval, but it doesn't have to be catastrophic – it could actually be good. The experience of migrating and seeing a society – your new home – through the eyes of another culture can be a creative trigger. The music, cuisine, languages spoken . . . all multiply through immigration, and this diversity greatly enriches nations, producing more inclusive, tolerant and interesting cities. But there will also be some cultural losses – national customs and traditions that fall out of fashion, replaced by other innovations. Britain's most popular foods today include spaghetti bolognaise and chicken tikka masala; whereas some native cuisine, such as jellied eels, once a very popular dish in London, has fallen out of favour. My immigrant grand-mother, born in 1922, retained her extremely conservative palate in her adopted home for only foods that had been available during her Central European childhood. My grandfather savoured the foods of as many different nations as he could. Some people will thrive on the increase in diversity, others should be reassured that their familiar culture will remain available in some form for as long as there is desire for it. This transition is not an abrupt change from one culture to another, it is a cultural fusion in which different traditions and ideas blend and enrich each other.

Large-scale migration can work. At the time of writing, it's too early to say how successfully the EU will manage the 2022 influx of millions of refugees from destroyed cities in Ukraine. However, in the past three decades, around 400 million people in China have migrated to cities, and the immense building and infrastructure programmes that made this possible have transformed the country, which today is 60 per cent urban. The state managed this unprecedented city migration without the explosion of vast, poverty-ridden slums that has blighted urbaniz-ation elsewhere, largely through channelling its migrants into small or medium-sized cities, where there has been crowding but very few slums. It's done this partly though the controversial *hukou* system, which imposes strict limits on where households can migrate, and through devolution of public services and many administrative functions to city governments.[10] This gave local governments much-needed agency over where and how immigrating populations settle, enabling the rapid demographic shift to be managed effectively and with low unemployment. China has also been frugal about its use of land space for urban development – cities now occupy about 4.4 per cent of the total land area. China has shown it's possible to create large migrant cities rapidly, and this level of ambition will be required across the northern latitudes to accommodate the new global migration. We've all shown we can adapt rapidly and build quickly when we need to. In just nine days during the pandemic of 2020, London converted an empty building into a 4,000-patient hospital, while Wuhan in China built a 1,000-bed hospital from the bare ground in just ten days.

Our post-war institutions were built for an international world; now we live in a global world. By this, I mean that the modern world has been constructed around an idea of internationalism, which is essentially a dominant network of rich Western nations that support each other. This network will have to expand and this will be a challenge, but the consequences don't have to be catastrophic. We can make this new world a good place for humans and nature to thrive. Recognizing the challenge of limited habitable land and resources in a hotter, crowded world offers the chance for a critical reflection on how our history has resulted in a situation where a person's life chances are so greatly defined by the geography and politics of where they were born.

The coming upheaval gives us an opportunity to disrupt this inequity by recognizing and protecting the rights of all migrants as global citizens. It's a chance to acknowledge that we have more in common than we have differences. And if this seems unrealistic or impossible, consider the enormous social change that we all willingly undertook, and achieved, in a matter of weeks during the 2020 global pandemic. Much of that cooperation occurred in the absence of conflict or authoritarian leadership. Consider, also, the cooperation that nations undertook to develop drugs and vaccines, to share scientific data and health interventions. Consider that populations everywhere, along with big corporations such as Astra-Zeneca, and NGOs such as the Gates Foundation, insisted that any vaccine be also made available to the poorest of our global society and not be sequestered or locked away behind patents by any one company or rich state. Yes, there was resistance, and certainly there are huge inequities, yet within two years of the virus first being identified, a billion doses of protective vaccine had been distributed and received by well over half of the world's poorest people.

So, we've done it before; we can do it again: we can cooperate on a grand scale to save lives. Migration, as we've seen, is the mother and the child of cooperation.

REINVENTING THE NATION

Remember, cooperation has been more important in the evolution of humans than conflict. We are superlative cooperators, even if this, paradoxically, also pushes us into racism and tribalism. These are unprecedented times, though. No national security threat comes close to that posed to states by global climate change and its social repercussions, including massive migration. Heatwaves already kill more people than wars.

Never has our species' cooperative ability been so necessary; never will it be so tested. The scale of our crisis requires new global cooperation, including new international citizenship and global bodies for migration and for the biosphere – new authorities that are paid for by our taxes and to which nation states are accountable. The political theorist David Held argued that we have outgrown our national boundaries through increasing globalization, and now live in 'overlapping communities of fate' from where we should form a cosmopolitan democracy at a global level.[11] Currently, the United Nations has no executive powers over the nation state, but if we are to bring down global temperatures, reduce the concentration of carbon dioxide in the atmosphere and restore the world's biodiversity, there will need to be globally imposed limits and management. We need some sort of global governance with enforceable powers.

Underpinning this global governance, we also need strong states, because the tension between the desires and needs of the individual and society are very real for us all, and hard enough to reconcile when our society is a small, closely knit group, let alone the population of the whole planet. It's hard to care about a nameless, faceless stranger in a country you've never visited when making life choices in a city thousands of miles away – indeed it's hard to balance the needs of a stranger one street away. That is, after all, what successful nation states help manage with the structure and institutions that ensure a useful level of cooperation between strangers that nurtures a strong society in which we all can succeed. We are not actually so closely related to other members of our society that cooperating with them makes sense genetically – but we cooperate with our group as though they were our family. We willingly make daily little sacrifices of time, energy and resources as individuals to ensure our society benefits – and we do this because it's *our* society, our social family, our nation state. The invention of the nation state has been a very powerful tool in enabling us to cooperate so well; as the political theorist David Miller put it, 'nations are communities that do things together.'[12]

Now, though, we need a blend between internationalism and nationalism. Only strong nation states will be capable of setting up the systems of governance that will help us survive climate change. Only strong nation states will be able to manage a massive influx of migrants from different geographies and cultures to the native population. In recent decades, the growth of globalization has led to greater internationalism – a citizen of London will often feel more commonality with a citizen of Amsterdam or Taiwan than with someone from a small country town in Britain. This may not matter for many successful urbanites, but natives of more rural areas can feel a loss. People need to belong, and with the decline of large industries and their unions, with the loss of social spaces and cultural traditions, many feel left behind by their own country. This creates resentment and fear of the kind that can lead to prejudice against immigrants. Liberalism, with its focus on individual autonomy, has failed to address this loss of national identity, leaving a vacancy for populist narratives and ideology to fill.

Instead, we must reinvent the nation state. We need it to be inclusive and not based on ancestry, skin colour or other divisive (and meaningless) characteristics. We need to feel commonality, a kinship with our fellows, based on our shared societal project, language and cultural works. Patriotism matters to people enough to make it a powerful source of identity. We could do worse than start with our nation's air, land and water, and the importance of defending them. We all face environmental threat, so enlisting military and other security institutions in the struggle against climate change is one ideological bridge. National service for younger citizens and immigrants to help with disaster relief, nature restoration, agricultural and social efforts would be another solidarity-creating step. And we need to restore or invent new national traditions that are environmentally or socially beneficial, and for which citizens can feel pride and respect. These could include social groups and clubs that sing, create, play sport or perform together, and to which members can belong for life. These traditions will help maintain dignity in hard times and provide patriotic meaning for immigrants to assimilate to. We need to strengthen our local connections while forging greater and more equitable global networks. The new patriotic narrative could be about civic nationalism, based on the common good, with rights and duties, and a passionate cultural attachment to nature, and to protecting and conserving places of national (or international) importance. The heroes we revere must be chosen to reflect the cosmopolitan nature of our society.

Costa Rica, for instance, embraced the term '*pura vida*', broadly meaning 'good life', as a national ethos, mantra and identity. Its use grew from the 1970s, when refugees from the violent conflicts in neighbouring Guatemala, Nicaragua and El Salvador poured into the country. Costa Rica, a small Central American country that has no standing army and instead invests in nature protection and restoration and social services, including health and education, used the phrase to help define its character and people to new immigrants. 'A person choosing to use this phrase thus is not only alluding to this shared ideology and identity, he/she is at the same time constructing that identity by means of expressing it,' explains Anna Marie Trester of New York University. 'Language is a very important tool of self-construction.'[13]

National pride doesn't have to mean seeing 'your people' as better than other nations', nor does it mean a centralization of meaning and power – instead, it can involve the devolution of traditions and an appreciation of regionality and of the enormous cultural value of new citizens. The European Union is an example of supranational identity that works because citizens can feel themselves to be European and identify with the values of the EU, but they don't have to give up their national identity or formally swear allegiance to some narrow historical definition of ethnic purity. Individual nations need to better apply the same idea. In the UK, for instance, London's Chinatown is rightly a much-visited tourist destination, as is Little India – they are part of the nation's identity, even though Chinese-Brits and British Indians often face prejudice and socioeconomic disadvantage.

To earn national pride rather than suffer divisive tribalism, a nation needs to reduce inequality. The state must invest in the people for the people to feel invested in the state. That means regulations and limits to free-market capitalism by applying social and environmental regulatory controls for the benefit of all, rather than simply a small tribe of global aristocrats. The Green New Deals proposed in the European Union and the US are examples of policies aimed at restoring economies and providing jobs and dignity while helping unite people in a bigger social project of environmental transformation.

Try, if you will, to clear from your mind the idea of people being fixed to a location they were born in, as if it affects your value as a person or your rights as an individual. As if nationality were anything more than an arbitrary line drawn on a map.

While we sit cosy in our safe homes, millions of migrants are desperately seeking the same: a chance to work, to contribute to a new society, and enjoy a dignified life for their family. Many are educated professionals who never dreamt they'd have to leave home. Increasingly we will be among them. There comes a point at which it is too expensive, difficult or dangerous to stay. When you cannot get wildfire insurance for your house, when it becomes too expensive to repair your home after yet another flood, when you realize you're spending most months of the year inside your house with the air-conditioning on because it's just too hot to go outside, when businesses and shops are boarded up and most of the houses are empty because the place where you live is no longer viable ... Then you may find yourself migrating somewhere else. Somewhere you can build a viable life for yourself and your family.

7

Haven Earth

Migration will remake the world in the coming century whether by accident or design. Far better the latter. Developing a radical plan for humanity to survive a 3–4°C-hotter world includes building vast new cities in the far north while abandoning huge areas of the tropics, and relying on new forms of agriculture. It involves adapting to a changed planet and our rapidly changing demography.

Our best hope lies in cooperating as never before: decoupling the political map from geography. However unrealistic it sounds, we need to look at the world afresh and develop new plans based on geology, geography and ecology – not politics. In other words, identify where the freshwater resources are, where the safe temperatures are, where gets the most solar or wind energy, and then plan population, food and energy production around that.

If we allow 20 square metres of space per person – more than double the minimum habitable space allowed per person under the English planning regulations, for example – 11 billion people would need 220,000 square kilometres of land to live on.[1] There should be plenty of room to house everyone on earth in a single country – the surface area of Canada alone is 9.9 million square kilometres. Of course, I'm not proposing anything as absurd, but this is something to reflect on when it is claimed that a country is 'too full' for any more people.

So let's look at the world anew, as benevolent Human Conservationists, and decide where are the best places for the relocation of this tricky species in the decades ahead.

The bad news is that no place on Earth will be unaffected by climate change. Everywhere will undergo some kind of transformation in response to changes in the climate, whether through direct impacts or the indirect result of being part of a globally interconnected biophysical and socioeconomic system. Extreme events are already occurring around the world and will continue to hit 'safe' places. Some places, though, will be more easily adaptable to these changes, while others will become entirely uninhabitable fairly quickly. By 2100 it will be a different planet, so let's focus on some of the liveable options.

Global heating is shifting the geographical position of our species' temperature niche northwards, and people will follow. The optimum climate for human productivity – the best conditions for both agricultural and non-agricultural output – turns out to be an average temperature of 11°C to 15°C, according to a 2020 study. This global niche is where human populations have concentrated for millennia, including for the entirety of human civilization, so it's unsurprising that our crops, livestock and other economic practices are ideally adapted to these conditions. The researchers show that, depending on scenarios of population growth and warming, '1 to 3 billion people are projected to be left outside the climate conditions that have served humanity well over the past 6,000 years.' They add that, 'in the absence of migration, one third of the global population is projected to experience mean average temperatures above 29°C, [which are] currently found in only 0.8 per cent of the Earth's land surface, mostly concentrated in the Sahara.'[2]

As a general rule, people will need to move away from the equator, and from coastlines, small islands (which will shrink in size) and arid or desert regions. Rainforests and woodlands are also places to avoid, due to fire risk. Populations are going to shift inland, towards lakes, higher elevations and northern latitudes. Looking at the globe, it is immediately clear that land is mainly distributed in the north – less than a third of Earth's land is in the southern hemisphere and most of that is either in the tropics or Antarctica. So the scope for climate migrants to seek refuge in the south is limited. Patagonia is the main option, although it is already suffering from droughts, but agriculture and settlement there will remain possible as the global temperature rises this century. The main lands of opportunity for migrants, however, are in the north. Temperatures in these safer regions will rise – and will rise faster in higher latitudes than at the equator – but the average absolute temperature will still be far lower than in the tropics. Of course climate disruption brings extreme weather, and nowhere will be spared these increasingly common events – Canada reached temperatures of 50°C in 2021, making British Columbia hotter than the Sahara Desert, then, a few months later, was hit by deadly floods and landslides that displaced thousands. Fires have blazed across Siberia's tundra, and melting permafrost is a shifting, unstable ground on which to build infrastructure.

Happily, however, the northern latitudes are already home to wealthier nations that generally have strong institutions and stable governments that are among the best placed to build social and technological resilience to the challenges this century. Problematically, many of them have also struggled politically with immigration to a far greater extent than have many much poorer countries (poor countries also host by far the greatest numbers of displaced people), and with a migrant 'crisis' that is far smaller than the great climate migration we will see over the next seventy-five years. It may be more possible to shift a political-social mindset in the space of a few years, however, than to return the tropics to habitability. Consider that most of Europe's nations each rely on tens of thousands of migrant workers just to harvest the crops they grow today. With better agricultural conditions across the north, the need for labour will only increase.

THE CITIES OF THE NEW NORTH

North of the 45°N parallel will be the twenty-first century's booming haven: it represents 15 per cent of the planet's area but holds 29 per cent of its ice-free land, and is currently home to a small fraction of the world's (ageing) people. It's also entering that optimum climate for human productivity with mean average temperatures of around 13°C.

Inland lake systems, like the Great Lakes region of Canada and the US, will see a huge influx of migrants – reversing the previous exodus from these areas – as the vast bodies of water should keep the region fairly temperate. Duluth in Minnesota on Lake Superior bills itself as the most climate-proof city in the United States, although it's already dealing with fluctuating water levels. Other upper Midwest cities around the lakes, including Minneapolis and Madison, are also likely to be desirable destinations. More southerly Midwestern cities face the threat of extreme heatwaves. The University of Notre Dame's Global Adaptation Initiative researchers concluded that 'eight of the top 10 cities facing the highest likelihood of extreme heat in 2040 are located in the Midwest', including cities from Detroit to Grand Rapids. Further east, locations get riskier quickly, but Buffalo in New York State, and Toronto and Ottawa in Canada look to be safer choices for migrants from the coasts.

Preparation and adaptation could enable some cities to survive on a coastal location. Boston, for instance, is far enough north to escape much of the projected extreme heat, and planners have developed a detailed strategy that includes elevating roads, building up coastal defences and introducing marshes to absorb flood waters. New York City, which faces extreme threats but might be too important to fail, is similarly planning extensive defences, although it's unclear how effective these will prove. Coastal cities that are far enough north and have steep enough coasts to protect against storm surges as sea levels rise will be safer.

Much of the rest of the US will be problematic for one reason or another. The central corridor will see worsening tornadoes; below the 42nd parallel, heatwaves, wildfires and drought will be perilous; at the coasts, flooding, erosion and freshwater fouling will be an issue. Today's desirable locations, such as Florida, California and Hawaii, will be increasingly deserted for the more pleasant climates of former Rustbelt cities that will experience a renaissance, as a globally diverse community of new immigrants revitalizes them.

Alaska looks the best place to live in the United States, though, and cities will need to be built to accommodate millions of migrants heading for the newly busy Anthropocene Arctic. In 2017, the US Environmental Protection Agency released a Climate Resilience Screening Index, which ranked Kodiak Island, Alaska, as being at the lowest risk of climate events in the country.[3] By 2047, Alaska could be experiencing average monthly temperatures similar to Florida today, according to an analysis of climate models.[4] As with everywhere, location is key, though – the residents of Newtok, Alaska, are relocating because melting permafrost and increasing erosion have caused portions of their village to wash away.[5] The retreat of ice sheets and melting of tundras is already causing huge problems for indigenous communities, whose way of life is being irrevocably altered. Their terrible loss, and that faced by native wildlife – not to mention other dangers, including unknown pathogens lurking in the currently frozen tundras, waiting to be exposed – will be countered by the vast opportunities for development in the New North. This is where many of the tropical migrants will create new homes during the turbulent twenty-first century, while humanity battles to restore a liveable globe. Whether self-governed indigenous communities will welcome this influx of southern migrants or reject what is the latest in a long history of often violent intrusions remains to be seen. However, people will move north and they will need to be accommodated.

With agriculture newly possible and a bustling North Sea Passage shipping route, the far north will be transformed. The melting of Greenland's ice sheet – the largest on Earth after Antarctica – will expose new areas for people to live, farm and mine minerals. Buried beneath the Arctic ice of Greenland, Russia, the US and Canada, there is also useful agricultural soil and land to build cities upon, giving rise to a hub of connected Arctic cities.

Nuuk is one such city set to grow rapidly over the coming decades. The capital of Greenland (an autonomous outpost of Denmark) sits just below the Arctic Circle, where the effects of climate change are obvious – residents already talk of the years 'back when it was cold'. A science station in Greenland's interior recorded a spectacular rise of almost 11°C from the 1991 to 2003 summer averages.[6] Fisheries here are experiencing a boost: less ice means boats can fish close to shore year round, while warmer ocean temperatures have drawn new fish species further north into Greenland's waters. Some halibut and cod have even increased in size, adding commercial value to fish catches. Land exposed by the retreating ice is opening up new farming opportunities with a longer growing season and plentiful irrigation. Nuuk's farmers are now harvesting new crops, including potatoes, radishes and broccoli. The retreating ice is also exposing mining opportunities and offshore exploration, including for oil. Nuuk stands at the edge of real economic gain. The country already has five hydroelectric plants to turn its abundant meltwater into power. According to projections Greenland will even have forests by 2100.[7] It may be among the best places to live.

Similarly, Canada, Siberia and other parts of Russia, Iceland, the Nordic nations and Scotland will all continue to see benefits from global heating, which was predicted over a century ago by one of the region's most celebrated scientists. In 1908, a decade after he had demonstrated that carbon dioxide emissions cause global heating, Swedish chemist Svante Arrhenius wrote that by burning fossil fuels 'we may hope to enjoy ages with more equable and better climates, especially as regards the colder regions of the earth, ages when the earth will bring forth much more abundant crops than at present, for the benefit of rapidly propagating mankind.'[8]

Arctic net primary productivity, which is the amount of vegetation that grows each year, will nearly double by the 2080s, with an end to cripplingly cold winters, according to projections that factor in the huge amplification of global warming around the poles.[9] (Winter average temperatures in the Arctic are already exceeding IPCC predictions for a 2°C temperature rise.) The Nordic nations already enjoy relatively warm temperatures because of the North Atlantic currents, but continental temperatures, which can plunge below −40°C in winter, will also ease, making interior locations more bearable.

Indeed, global heating has already boosted Sweden's per capita GDP by 25 per cent, a Stanford study found. The biggest greenhouse gas emitters 'enjoy on average about 10 per cent higher per capita GDP today than they would have in a world without warming, while the lowest emitters have been dragged down by about 25 per cent', the researchers found.[10] The moral argument for including tropical migrants in the economies of the north is clear. The researchers estimate that India's GDP per capita has lagged by 31 per cent owing to global heating; Nigeria's has lagged by 29 per cent; Indonesia's by 27 per cent; and Brazil's by 25 per cent. Together, those four countries hold about a quarter of the world's population – at global temperature rises of 2–4°C, the economic impact will be far more crippling. Those African, Latin American, South and Southeast Asian inhabitants will need to take their place in the habitable far north.

Nordic nations score comparatively low on climate change

vulnerability and high on adaptive readiness. Growing seasons will significantly expand, particularly around today's farmland,[11] with new species of plants and animals thriving. Forestry cultivation alone could increase by one-third in the north, according to studies. The birch forest is already advancing north at 40–50 metres a year, transforming ecosystems and melting permafrost.[12] And the region's use of electricity is projected to fall the most in Europe as warming winters reduce the demand for heating.[13]

In a picturesque location near Europe's largest glacier, the small coastal town of Höfn in southeastern Iceland could be another winner. The town today is reliant on fishing, especially of lobster, and tourism, with scope to diversify and expand. Iceland is actually experiencing falling sea levels, because as its weighty glaciers pressing down on the continental plate melt, the land is rising back up. Scotland, too, is still rebounding after the end of the last ice age. Currently, the southeast coast of Iceland is rising particularly fast, and as commercial traffic increases through the newly ice-free Northwest Passage – the sea route through the Arctic connecting the Atlantic and Pacific Oceans – Höfn's harbour stands to win. To be clear, the Arctic Ocean will continue to freeze in winter, but rapid ice melt will make the Northwest Passage open and navigable for shipping for much of the year, cutting shipping times by around 40 per cent. This will enable easier regional trade, tourism, fishing and travel, as well as open opportunities for mineral exploration.

Another handily placed port town, Churchill in Manitoba, Canada, will also profit from climate change. This barren outpost, wedged between boreal forest, Arctic tundra and Hudson Bay, has just 1,100 residents, who rely almost entirely on polar-bear tourism. Churchill's land was considered so undesirable that in 1990 the US freight company OmniTrax bought the town's port from the Canadian government for $7. However, with an active migration programme recruiting people and businesses from around the world, the newly developed city could support international trade through its revitalized port on the Hudson Bay – the only commercial deep-water port in northern Canada. This could make it a key stopping and unloading point on the Northwest Passage for cargo ships coming all the way from Shanghai. Churchill is connected to Winnipeg and the rest of

Canada – and the US – via its restored railway line. And it's just over 100 kilometres from Nunavet, Canada's newest indigenous province, a growing Inuit-governed territory. The tourism industry here is world-leading, water is plentiful, and the winters warmer.

Churchill could become a booming city. Indeed, Canada will be a key destination for our migrants, and the government is betting on it, aiming to triple the population by 2100. The nation is currently adding 400,000 new immigrants a year with the aim of increasing the population from 37 to 100 million. 'Canada is built on immigration, and we will continue to safely welcome the immigrants that Canada needs to succeed,' said immigration minister Sean Fraser, in December 2021. 'I can't wait to see the incredible contributions that our 401,000 new neighbours make in communities across the country.'[14]

Most of Canada's recent immigrants have been attracted from climate-threatened nations in Asia, with India, China and the Philippines dominating. Marshall Burke, Deputy Director of the Center on Food Security and the Environment at Stanford University, calculated that global heating could raise the average income in Canada by 250 per cent due to greatly expanded growing seasons, reduced infrastructure costs and increased maritime shipping.[15] With a stable, non-corrupt democracy, one-fifth of the world's freshwater reserves and as much as 4.2 million square kilometres of newly arable farmland, Canada could be the world's new breadbasket later this century.

Russia will be another net winner this century – its 2020 national action plan explicitly describes ways to 'use the advantages' of climate warming. According to the US National Intelligence Council, Russia 'has the potential to gain the most from increasingly temperate weather'. The country, which spreads over 10 per cent of the planet's landmass, is already the world's biggest exporter of wheat, and its agricultural dominance is set to grow as its climate improves. A country that can feed the world can wield increasing power over the world, no matter the political landscape today. Retreating permafrost, which holds the world's greatest concentration of soil carbon, is exposing nutrient-rich virgin land for cultivation, but also risks catastrophic carbon release. By 2080, more than half of Siberia's permafrost will have gone, and the most inhospitable third of the massive Eurasian nation will switch from 'absolute extreme' to 'fairly favourable' for civilization, according to a detailed modelling study.[16] Climate-wise, the frozen north will become more attractive and able to support much larger populations. Longer growing seasons across the north and east of Russia will mean a particularly productive future for places like Yakutsk, the capital city of the Sakha Republic. Yakutsk is already a major supplier of diamonds, and has large reserves of gold and other minerals. The retreat of the Siberian permafrost zone will prove an even greater boom for mineral extraction – by 2050, permafrost could have retreated by more than 100 miles.

Though there is much potential gain, the loss of permafrost and of ice roads will be hugely problematic for many settlements – including large cities – in the continental interior of Russia and, to a lesser extent, Canada. Essentially a bog that has frozen solid, permafrost makes a good foundation for buildings, roads, railway tracks and other infrastructure. But not once it melts back to bog. Nearly 70 per cent of current infrastructure built on permafrost is at high risk of surface thaw by 2050, according to an assessment in 2022.[17] There are effective engineering techniques to deal with the problem, but they are expensive, costing more than $35bn a year by 2060, the researchers estimate. Canada's sparsely occupied Northwest Territories already spend around $41m a year on permafrost damage, which is about $900 per resident.[18]

Many towns in Siberia, built during Stalin's gulag programme, are inaccessible except by air or during the few months of each year when ice roads can be created. Once it is too warm for ice roads, these places will effectively be isolated. The interior of Russia will likely face the same emigration as climate-hit places of lower latitudes, whereas the coastal cities will be booming. However, in time, these thawed bogs will stabilize and the landscape will be exploitable for drainage, construction or agriculture. Even Russia, which is depopulating at its fastest rate (it lost almost 1 million people in the year 2020–21, yet has been too xenophobic to embrace migrants[19]), is changing tack. The state recognizes that, without a growing population, Russia risks losing its already diminished geopolitical clout as well as declining in economic strength. With wild lands across Russia's east able to be transformed into farmland for the first time, a migrant workforce, largely from China, is already growing wheat, corn and soya there. In 2020, Vladimir Putin allowed dual citizenship, hoping to convince its immigrants to become Russian. However, economic sanctions triggered by its aggressive military operations do little to enhance Russia's desirability.

Other places that will see new or expanded cities include Scotland, Ireland, Estonia, and elevated sites with plenty of water, like Carcassonne in France, which is surrounded by rivers. In the global south, as mentioned, there is far less landmass in the high latitudes, but Patagonia, Tasmania and New Zealand, and perhaps the newly ice-free parts of the western Antarctic coast, offer potential for cities. In Antarctica alone, up to 17,000 square kilometres of new, ice-free land is projected to appear by the end of the century. This could offer an opportunity for development, but I fervently hope that Earth's last wild continent will remain a precious nature reserve.

Elsewhere, people will move to higher elevations, although these too are becoming hotter, especially as they lose ice and, with it, the availability of fresh water. High-altitude locations that will see an influx of migrants include the Rocky Mountains in North America and the Alps in Europe. Switzerland, for instance, has lakes and altitude. In the US, Boulder and Denver, both above 1,600 metres, are already attracting migrants, and Ljubljana in Slovenia is another alpine location with a rich underground aquifer system and lush agriculture.

Many mountain cities are already refuges for climate migrants at 1.2°C, but won't be suitable at higher temperatures. Medellín in Colombia has plenty of freshwater surrounded by fertile agricultural land, and has drawn thousands of people from drier, harsher parts of the country. However, its tropical location means it will be hit hard by climate change, in particular by fiercer and more damaging rainstorms that create landslides and flooding, and risk structural collapse. The city has been working to improve the resilience of its infrastructure, but as the region suffers further climate shock its fragile social system is also at risk. Colombia has undergone decades of insurgency and violence, and remains very poor. Like most of Latin America, it is likely to become a generator rather than receiver of migrants.

People will aim for safer places, and they will be better off moving to locations that already have good governance, productivity and resources. Happily, there are many places where these coincide. Some of this migration will involve rapidly expanding existing towns and cities; in other places, such as Russian Siberia and Greenland, entirely new cities will need to be built.

OPENING BORDERS

Finding a suitable place for relocation is only the first part of the job for human migration, though. Unlike animals, or our footloose ancestors who could simply pack up their tents and move, we are all part of a complex social mesh that can become a cage. For human migration, particularly an upheaval on this unprecedented scale, we need to look at how this would work in terms of territories, borders and the twenty-first century world we've made.

Achieving safe settlement for hundreds of millions of migrants could require the compulsory purchase by international consensus of land held by current states, with compensation and a stake in the new cities and their industries. It could require a new kind of international citizenship. It could mean richer, safer-latitude states becoming 'caretaker states' for poorer, more vulnerable ones, during the crisis period of global heating until planetary restoration. It could involve charter cities, states within states, the extinction of some of the 200 nation states and consolidation of the remaining few into regional geopolitical entities. There are many alternative visions to today's status quo of nation states, borders and passports – which are, after all, relatively recent.

Instituting global freedom of movement, for instance, would boost national economies, as well as saving or improving billions of lives. Open borders would, it's fair to assume, result in very large flows of people – estimations range from a few million to more than 1 billion – and could increase global GDP by tens of trillions of dollars.

However, opening borders doesn't have to mean no borders or the abolition of the nation state. In the brief time we have to prepare for massive disruption to our planetary system, it seems unwise to try also to completely abandon our geopolitical system. After all, a big part of the desirability of the migratory destinations we've selected is the result of the functionality of the nation state: the institutions, rule of law, investment decisions, infrastructure and other policies that have led them to be thriving places capable of attracting and keeping migrants. Workers in rich countries earn more partly because they live in societies that have developed institutions that foster peace and prosperity. Bluntly, some countries are better run than others.

Given that significant populations of some countries, such as Bangladeshis or Vietnamese, will be arriving in another nation, where in some instances they may outnumber the native population, they must not simply be absorbed into native political structures, but given representation. Done carefully and sensitively, through intelligent legal structures, this should enable immigrants to feel valued and to maintain their dignity, while also ensuring natives don't feel pushed out or overwhelmed. Native-born Canadians will be outnumbered 2:1 by immigrants by 2100, for instance. Key to Canada's success will be the maintenance of its existing political social systems, but equally important will be the recognition of the specific social and cultural needs of its newest citizens.

There are also alternatives to the nation-based geopolitical system we have inherited, such as small but powerful city-states, which were the norm in Ancient Greece and Renaissance Italy, and today include Singapore and, less explicitly, Dubai, Macau and Hong Kong. In the coming decades, large megacities could enjoy increased autonomy, including over their labour, border and visa policies, and become effectively city-states, even if they remain part of larger nations for other functions. Another option would be new regional unions with free movement of trade and labour, and governing powers. The European Union is a successful example that uses a single currency and has limited governing power. Other blocs, such as the African Union, are following suit. In future decades, Arctic nations including the Nordic countries, Greenland, Iceland and Canada may well form such a union, with agreed policies on managing immigrations, the shared ecosystem, mineral extraction and shipping, for instance. As the climate shifts and migration increases, it makes sense for nations that share borders to strengthen their relationships and find common solutions around issues from labour and goods to energy and resources, gaining shared resilience.

CHARTER CITIES

Another option might be charter cities, set up and operating under a different set of rules to the jurisdiction they are next to. Charter cities were originally proposed by Nobel Prize-winning economist Paul Romer in 2009, as a new type of governance structure to boost development in poor nations. According to his plan, poor countries would donate territory to a wealthy, well-run nation, such as Switzerland, to govern effectively. Romer's idea is that the citizens of the charter city would benefit from good governance, safety and wealth; the host nation would receive taxes, plus the benefits of having a well-developed economic hub in their country; and the governing nation would get investment opportunities and comparatively cheap labour and resources.

The idea isn't so far from the concept of a 'special economic zone', which rapidly transformed cities including Shenzhen in China and Dubai in the United Arab Emirates. Essentially, these are areas of a nation that operate special business policies and laws with the aim of attracting foreign investment, and increasing trade and employment. Singapore and Hong Kong are similar success stories that became rich on the back of better legal systems, less corruption, stronger rule of law and more competent administration. Notably, Singapore and Hong Kong also benefit from their strategic locations: both are gateway cities that control massive trade flows through the Malacca Straits and the Pearl River Delta respectively.

Romer's model wouldn't be suitable for the needs of the coming decades, of course, as large populations will be migrating away from climate-ravaged lands – and the idea that poor countries exchange sovereignty for a hike out of poverty is likely to prove unpalatable for many. Nevertheless, private charter cities are already under way, mooted as an economic development model that enables residents to escape climate impacts. Honduras has created an embryonic charter city called Próspera ZEDE (the Spanish acronym for 'Zone for Employment and Economic Development') on a small, 58-acre plot of unoccupied land on its Caribbean island of Roatan. So far, there are just three buildings, but it plans to expand to 10,000 residents by

2025, who will need only to sign a social contract and pay a considerable membership fee, to be a part of the libertarian dream.[20]

The concept shares elements with the Seasteading movement, a libertarian group of mega-rich preppers intent on building independent floating cities on the high seas. The Seasteading Institute was founded in San Francisco in 2008 by anarcho-capitalist (and Google software engineer) Patri Friedman, with funding from PayPal billionaire Peter Thiel, to 'establish permanent, autonomous ocean communities to enable experimentation and innovation with diverse social, political, and legal systems'. Some of the ideas they plan to use include harvesting calcium carbonate from seawater to create 3D-printed 'artificial coral' cities of upside-down skyscrapers – 'seascrapers' – powered by oceanic geothermal energy. Some of this energy will be used to draw nutrients from deeper waters to the surface to grow seaweeds in farms worked on by 'the poorest billion people on earth', welcomed because 'floating societies will require refugees to survive economically'. These floating utopias will 'liberate humanity from politicians' while solving the planet's big problems, it is claimed. For the more sceptical among us, this smells dystopian, rather.

The group's first venture, a floating charter city in French Polynesia, stalled at the planning stage due to negative reaction from the French Polynesians, concerned about pollution, disruption and environmental damage. One Tahitian TV host compared the situation to the evil Galactic Empire in *Star Wars* imposing on the innocent Ewoks, while secretly building the Death Star.[21]

Undeterred, Bitcoin tycoon Chad Elwartowski and his partner, Supranee Thepdet, built a Seastead cabin off the coast of Phuket in Thailand in 2019. However, the Thai government charged them with violating national sovereignty, which carries the death penalty, and the couple fled, escaping Thai naval police by minutes. They are now working on another Seasteading project, through their newly formed company Ocean Builders, off the Caribbean coast of Panama, this time with the agreement of the government of Panama.

Even if the Panamanian venture or the Honduran charter city prove successful, their locations in the tropical Caribbean are far from ideal given their vulnerability to climate change – however, with enough money and engineering, it's possible that a small population could make a survivable home in these locations. The more important message from these examples is the window they offer into what our unplanned migratory future could be: one in which small groups of the wealthy elite use their money and power to prepare liveable islands of isolation, while hundreds of millions of people are stranded in the deadly environment that they themselves did little to create. That scenario has been explored in plenty of sci-fi depictions so there is no excuse for a failure of imagination – it's up to us to ensure it stays as science fiction. We must plan today for population-scale survival with the same level of ambition as shown by the wealthy few prepping their retreats.

In 2015, at the height of the Syrian refugee crisis, Egyptian billionaire Naguib Sawiris offered to buy several small Greek islands, each of which could be used to house around 30,000 refugees. Sawiris compiled a list of twenty-three private Greek islands, owned by investors willing to sell to him, and took it to Greek Prime Minister Alexis Tsipras and the United Nations refugee agency UNHCR. His proposal was that the refugees would help build temporary developments, which would eventually be used for tourism once the Syrian conflict was over. Sawiris planned a joint stock company with US$100 million in capital that would receive public donations. To date, nothing has happened with this project, but as a temporary response to extreme events that generate sudden waves of migration, a combination of private investors and public support could be used to create refuges on private islands. In the longer term, however, people cannot make lives on isolated islands and must be accommodated in wider society.

There may well be a role for charter cities, but not in the hazardous equatorial belt, as originally conceived. Instead, they would need to be built at higher latitudes and focus on much larger populations moving to satellite cities, which could be operated by the origin state but located on land, rented or owned, within another state. Charter cities of this model may well be a good solution for states such as Nigeria, Bangladesh or Maldives that could buy or rent land inside a large country such as Canada, Russia or Greenland, effectively gaining habitable territory, for the period of ninety-nine years, say. Consider that little Nigeria has a population bigger than Russia and Canada combined (it will reach 400 million by 2050), whereas these two countries have vast amounts of land. The rented administration would be run by its mother nation, and perhaps 'taxed' by the land-lord nation through produce. Once the rental period was up, the lease could be extended, or the territory returned to its former owners. Citizens could be given the option of remaining as citizens of their new nation or relocating to their mother nation, which, given decades of climate restoration, would hopefully be safe enough to inhabit.

This may sound far-fetched, but purchasing or leasing territory has occurred in other places – Britain leased Hong Kong from China for ninety-nine years; the US purchased much of its territory from other nations, including buying Alaska from Russia in 1867. As increasing areas become unliveable, states will face hard choices about how to manage the massive populations seeking safe locations – many people will be absorbed into existing political entities in the global north, thus expanding the power and productivity of states such as Canada and Russia; others may well find the translocation of existing nation states a more equitable and desirable solution, and seek to make it happen in a reversal of the nineteenth-century colonial expansions (ideally, this time through rental or purchase of territory, rather than enforced takeover). In 2014, the small Pacific island nation of Kiri-bati, which is becoming unliveable owing to rising sea levels, purchased 20 square kilometres of jungle territory in Fiji, which President Anote Tong said would initially be used for agriculture. 'We hope not to have to move everyone on to this one piece of land, but if it became absolutely necessary, we could do it,' he said.

One potential location for charter cities would be northeastern Russia, a massive region of depopulated cities with agricultural poten-tial and plenty of mineral resources. A nation like India or Bangladesh could lease land for a charter city, protecting Russia's sovereignty from its fear of Chinese encroachment, and the revived productivity would provide useful tax revenue. The depopulating nation favours immigrants from its former Soviet neighbours. However, the far east-ern region is newly awash with Chinese money – perhaps another outcome is that northeastern Russia becomes once again Chinese Manchuria, and China brings in migrants from across its Silk Road initiative to populate the region.

GIVING PEOPLE A PUSH

The biggest migration problem is not that there's too much of it, but that, even within national borders, there is simply not enough of it.

The rationale for migration this century is clear: it is essential to help us survive environmental change, poverty and global inequality. But not enough of us want to move. Encouraging and assisting migration needs to be a policy priority. Research suggests this is best done not by forcing or incentivizing people to move, but by removing obstacles.

The majority of people who live somewhere other than where they were born, moved through what demographers refer to as 'household formation'. In other words, someone left their parents' home and then made a new home with someone else in a different location. Multi-generational households have significantly declined in the West in recent decades, so this is very common. People create their new households in places that hold attractive opportunities, where there are jobs available or they can study, for instance. Economic differences lead to population shifts – migration – that occur over a period of years to generations from poor to rich places. The past few decades in the UK have seen a migratory drift towards the southeast, for example. In smaller countries, much of this occurs across borders, but in the largest nations, such as China and India, this large movement has mostly been internal. Only 3.5 per cent of the world's population are international migrants. The biggest international flows today are across the Mexican border to the US, and the Bangladesh border to India (although much is undocumented).

Climate change this century is going to alter economic geography – the relative attractiveness of a place to live. Most migratory movement is for work; the majority of migrants are working-age adults. The EU is one of the most popular destinations. Even then, only a small proportion of the population of Europe are immigrants. When Europeans talk about the 'swarms' of African migrants flooding their shores, it's worth noting that just 2.5 per cent of Africans live abroad, and less than half of these people actually leave the continent. Even though they would likely earn more, very few move. Part of the problem is, of

course, border restrictions, which mean that it is now extremely difficult and expensive to move from a poor to rich nation unless you
have professional skills that are highly valued or you are a close relative of a legal migrant. But this is only a part of the story – look at
movement within the European Union, which is effectively borderless
for member states. Wages are at least twice as high in Germany as in
Greece, and Greeks are free to move to Germany whenever they wish,
yet only around 150,000 out of a population of 11 million people
have done so in the past decade – just over 1 per cent. The language,
food and other aspects of German culture are very foreign for a Greek,
and being a foreigner is a social challenge, at least until a person forms
a network. It's much easier to stay.

Even when the income disparity is very much greater, people are
sticky. Micronesians mostly stay where they were born, even though
they are free to live and work in the US without a visa, where the
average income is twenty times higher. Niger, next to Nigeria, is not
depopulated even though it is six times poorer and there are no border
controls between the countries. People like to stay in the communities
they were born in, where everything is familiar and easy, and many
require a substantial push to migrate – even to another location in the
same nation, and even when it would be obviously beneficial. One
study in Bangladesh found that a programme that offered subsidies to
help rural people migrate to the city for work during the lean season
didn't work, even when workers could make substantially more money
through seasonal migration.[22]

One problem is the lack of affordable housing and other facilities
in cities, meaning people end up living illegally in cramped, unregu-
lated spaces or in tents. Another is family life – childcare can be
unaffordable or unavailable in the city, whereas extended family can
provide this service for free in the village. Offering housing assistance
through rent subsidies, matching people to guaranteed jobs (however
temporary) pre-migration, and ensuring childcare support would all
help people make the move. This isn't just a problem in the poor
world, it also affects the person moving from small-town Michigan to
Chicago, say, where prospects are better, but housing prices are much
higher and childcare away from family is no longer free.

There is also a psychological reluctance to make risky decisions: if you make a decision to move and it doesn't turn out well, you'll be more disappointed than if you do nothing and it turns out badly. Offering migrants some insurance against failure can help tip the balance – when this was offered in the Bangladesh study through risk-sharing, the effects were almost as large as for the offer of a free bus ticket, with around 20 per cent more people making the move.[23]

People will put up with significant financial drawbacks and future uncertainty in exchange for pleasant conditions today. For instance, wages are average in Hawaii, the cost of living is exceptionally high, and the archipelago will become unliveable within decades. But lots of people move there anyway, because they enjoy the weather and lifestyle.[24] When I was in the Florida Keys recently, it was obvious that time was running out for the sun-drenched idyll – water was coming up through cracks in the streets, and parts of Key Largo were already submerged. Yet there was no shortage of realtors selling expensive property there. Consider the people of New Orleans, who saw rising sea levels and flooding for years before Hurricane Katrina. Research since that disaster has shown that survivors who migrated after the storm ended up earning more in their new cities.[25] So why didn't they move sooner? Because the benefits of migration only tipped the balance against the loss of leaving familiarity and their social circle once Katrina made it near impossible to stay.

If we want people to make the move safely, before disaster forces them, we're going to need to help them take the decision earlier.

It isn't just a case of missing your mates or being reluctant to leave grandma's cooking: as we've seen, humans are entirely reliant for survival on our social network. Moving to a new place removes you from the social connections that help people cope during emergencies, ill health, new motherhood, unemployment or mental health crises. The first refugees that fled Ukraine for safety were those who had family in their destination countries. In a very real way, we are reliant on each other, and for a migration to succeed, people have to be able to stretch their networks to span geographical distance, or forge new networks quickly. This is why state-sponsored support for migration is essential for migrants and an invaluable investment for host nations hoping to boost their economies with immigrants.

This means ensuring as never before that housing needs are met;

building infrastructure on a grand scale; giving people a subsistence allowance, training, opportunities to find work and create their own businesses; subsidizing childcare; paying people for caring roles; and helping them integrate and set up their new lives with language and citizenship classes. Punitive restrictions on social support for new migrants are not just inhumane but a false economy. Residency visas must be more flexible, giving immigrants several years to find jobs rather than brief periods, and without stipulating minimum earnings – experienced care-providers, home-help assistants, labourers, food preparers and delivery drivers are all vital roles but don't necessarily pay above visa threshold levels. This immediate, short-term support will be more than repaid over the longer term. However, today, many rich countries are failing to meet these needs even for their native populations, owing to poor social policy that has enabled soaring inequality. Nations will only meet the extraordinary challenges we face this century through bigger government support: strong regulations, public ownership of essential services, and policies that nurture communities – and businesses – systemically, by meeting the needs of workers, children, the sick, the elderly, and their environment. The additional costs of providing this social care for migrants would be proportionately small, whereas the future contributions of these integrated populations is potentially enormous. In other words, it's a safe investment.

8

Migrant Homes

Every year, the Peruvian capital Lima grows larger as one-room shacks expand over the surrounding desert. Instead of facilitating the rural poor to move to planned settlements with infrastructure and social assistance, the government largely ignores the 'problem' of migrants. In Lima's slums, as in shanty towns across the world, people pay to live on what becomes tightly controlled, gang-run land. Internal people-traffickers locate land on the outskirts of the city, ascertain that the owner is overseas, and then collect people to invade it – for a fee.

Back in 2012, I spoke to one of the slum's residents, Abel Cruz, a farmer from Echarate near Cusco, who had kept pigs and grown cocoa and vegetables. He told me that continued terrible drought had made life on the farm increasingly desperate.

'One day, a man came into the village and asked if we wanted a better life in Lima with good jobs and school for our two boys,' Cruz said. 'We didn't want to leave our families and our home, but the drought got worse.' Like millions of people across the tropics, Cruz decided the hardship and uncertainty of migration would be better than a future of hopeless poverty and hunger in his village.

'We were told to meet the man at 5 a.m. with the money and some sheets of bamboo ply. When we met him, there were lots of other families just like us there. We were taken to a sand dune and told to fence off an area and start constructing our houses from the material we brought with us.' Their home, like their neighbours', had a bare earth floor and few possessions. Rolled-up bed mats stood against one ply wall; a nest of bowls and a couple of pans were neatly stored in the corner.

'Our toilet is a hole we dig in the floor of the room and cover with plywood. Every couple of years it gets full, so we make another hole. It'll take twenty years for the whole of the floor-space to be full of shit, so I don't know what we'll do then. The place stinks and people get sick. We have no water and spend half our wages on water delivery from a truck.'

Initially, urban migrants in the global south usually move between the city and their village, returning to help with harvesting and at other labour-intensive periods, or when work dries up in the city. Seasonal migrants often sleep in makeshift dorms or camp on the floor of their workplace, saving every penny they earn for food or to send back home. Keeping one foot in each place enables them to build up networks of security and useful skills in the city while not losing out on any potential land ownership in their village. Eventually, the move to the city will become permanent, and then migrants build a life that may involve further migration to another city or country, learning new skills and exploiting new opportunities. The remittances, however, continue to flow back to their original rural village, and the networks that the migrants create and strengthen enable other villagers to make the same journey.

Migrants will keep coming to cities. Helping them transition quickly and painlessly would benefit everyone. That means doing something that few governments are good at: preparing for a crisis that hasn't yet happened.

In 2008, for the first time in our evolutionary history, more people lived in cities than on the land. We had become a different animal, living geographically divorced from the natural world that feeds and fuels us. The urban migration began in the West – between 1850 and 1910, as many as 2 million people a year moved to cities, and rural parts of the United States are consequently littered with ghost towns. The same is now occurring in developing nations as rural areas depopulate. It is the century of megacities.

Today, roughly one-third of humanity is on the move, mostly as internal migrants shifting from rural landscapes to cities. Urban migration is shaking up human geography as refugees flee environmental degradation and its social consequences. The challenge over the next eight decades is to create vast, densely populated cities that are safe, liveable and inclusive, with an economy that works for the

Anthropocene – recirculating water and resources, and managing waste and production, without polluting the natural environment.

The urban transition has largely been completed in the Americas and Western nations. In Asia, urbanization has been under way for some decades, led by Malaysia, China and Thailand, although there is some way to go, particularly in South Asia. In Africa, the population is still largely rural, but the urban population is increasing rapidly, at 3.6 per cent per year. This migration, in addition to the still-high birth rate, means Africa's cities will be home to 20 million extra people per year. This decade, the world's ten fastest growing cities will all be in Africa. Tanzania's Dar es Salaam, just a fishing village as recently as the nineteenth century, will be home to 11 million people by 2030, double today's population. Meanwhile, Lagos and Cairo are projected to be home to more than 24 million people each by 2030. None of them will be viable within a few decades of global heating – indeed all of these cities are already experiencing fatal impacts from extreme heat, flooding or both.

The massive migration to the north will be a transition of urbanites. International migration is usually made from cities, even if the migrants were born in a village.

Ideally, this urban growth will be matched by improvements to well-being, as occurred in Singapore post-independence, for instance. However, the urban transition under way in Africa, although more rapid than in Asia or Latin America, is happening at greater levels of poverty. It means that in Lagos, for example, city infrastructure isn't developing at anywhere near the pace needed to manage the new migrants, leading to vast, sprawling slums, bisected with narrow thoroughfares, poor sewerage, frequent power cuts and other problems. Nigeria gained independence from Britain the same year as Singapore, in 1960, but Lagos has grown faster and remains far poorer, with millions living in flood-prone, swampy land without electricity or sanitation, and with poor health and low literacy. Because people are spread out, rather than concentrated in planned districts, it makes business, trade, wealth creation and innovation far less efficient, and so the country far less productive. We will need to bear this in mind when planning future cities.

Cities work best when they are concentrated hubs. Although rural

Africans earn more when they migrate to the city, they don't earn a lot more, because unlike in other parts of the world, they don't easily end up in more productive jobs. They don't live near these jobs, so they must commute, which is expensive and takes a long time on jammed, narrow roads. Nairobi has one of the longest commuting times in the world because more than four out of ten people walk to work. This means that migrants end up working as street hawkers, selling vegetables or trinkets on the side of the road, rather than getting a more lucrative job. Poor infrastructure and city planning throughout the continent mean transporting anything is more expensive, pushing up the prices of food and other resources, which means factory wages and other costs are higher. That makes it harder to compete on the global market, so African cities are largely 'consumption cities' with economies dominated by low-value services and goods that are consumed locally rather than traded. These reasons help keep urban Africans significantly poorer than their counterparts on other continents. This makes them more vulnerable to the effects of climate change and other shocks and stresses – and a priority for mass migration.

There are multiple factors behind this, including the legacy of colonial exploitation, HIV, conflict and poor governance, plus an unproductive agricultural sector that makes food costly and limits incomes. However, better urban planning, with denser housing, wider roads, and good transport and infrastructure, would vastly improve the productivity and wealth of cities in Africa during the transition to urbanization this century – and improve people's climate resilience.

It's been a similar story, to a greater or lesser degree, across the world. Urban migration tends to be unplanned and iterative. The fanciest cities in the world began as commercial and administrative centres ringed by housing that became less salubrious and more slum-like as you progressed further out. The overcrowded shanty towns of eighteenth- and nineteenth-century Europe, home to the migrants that made their cities wealthy, were notoriously filthy death traps, plagued by typhoid, cholera, dysentery, malaria and other diseases of poor sanitation. In the winter, families froze; in summer, they festered and stank. Bit by bit these slums were knocked down as a result of twentieth-century development policies, and replaced with decent buildings and essential infrastructure. One of the most deadly slums of nineteenth-century London, Seven Dials, is now the upmarket heart of the city's Covent Garden theatre district. Similarly, the internationally notorious Five Points slum of New York is now desirable real estate between Manhattan's Chinatown and the Civic Center.

Cities that experience rapid urban migration generally expand through rural migrants 'illegally' building inexpensive shacks in clusters on the outskirts of the established city. As these grow in number, they start to envelop rural villages, which become new suburbs of the city. This kind of low-rise, land-hungry expansion is often inefficient, and can trap people in poverty – governments regularly neglect these areas because the housing is illegal, so the residents don't receive sanitation, water, health services and other essentials. They also live in constant fear of eviction, of returning from work to find their homes bulldozed, usually without compensation. However, the networks that migrants create and maintain, and the vital social capital that they bring, means these informal, affordable housing projects are usually vibrant, commercial markets, and an essential social-mobility path out of poverty. Slums, then, act as dynamic stepping-stones from rural poverty to an urban life of hope and opportunity.

Spitalfields, a district in east London, was once home to Huguenot immigrants from France, who became prosperous through silk weaving. By the early nineteenth century, competition from Manchester textile factories had driven the community into terrible poverty and their spacious houses had been subdivided into tiny, overcrowded slum dwellings. As the century progressed, the slum became home to Dutch and German Jews, then later to masses of poor Polish and Russian Jews, and other Eastern European immigrants. Twentieth-century social programmes improved the area's conditions, although it remained poor, and it became home to waves of Jewish then Irish immigrants and, towards the end of the century, to Bengali and Bangladeshi immigrants. By the turn of the twenty-first century, 'Banglatown', as the area was called, encompassing Shoreditch and Brick Lane, had become trendy for artists and other creatives.

There is usually a divide between the established city and the slum: slum dwellers are often denigrated as a group and thought of as a problem – criminals and dirty – even though the wealthier citizens – many themselves second- or third-generation migrants – rely on their services as domestic workers and in construction and other essential jobs. The two worlds, situated cheek-by-jowl, are separated by the hierarchy of circumstance – many wealthier residents will never set foot in a slum area on their doorstep.

Eventually, these slums, shanty towns, favelas, urban villages and so on become legitimized as part of the city proper, and their structures are made permanent. The irony is that the diversity of culture and entrepreneurship that characterizes these areas often makes them some of the most desirable parts of the city, which pushes up prices to the extent that the original community can no longer afford to live there once it has been gentrified. The former Spitalfields slum of Shoreditch is now among the most gentrified districts in London and, with the exception of a few social housing projects, far too expensive for the 'Bangla' communities. Poor people then get pushed further out to high-rise housing that has been built specifically for this purpose, but without the established social network or opportunity for entrepreneurship that gives a slum city its economically empowering dynamic. The result is people then become trapped in poverty in inner-city enclaves with little prospect of moving on.

ANTHROPOCENE CITIES

This, then, is our challenge. We need to make our migrant cities places of hope, where poor people are able to make their home and build strong networks, with opportunities for employment, training, entrepreneurship, or some other investment of their time and social capital. These cities must be safe and healthy with decent infrastructure. They must be affordable, ideally use no more electricity or water than they generate themselves, not contribute greenhouse gas emissions, and not worsen biodiversity loss. In as many ways as possible, material resources should circulate within their economies, minimizing waste and avoiding pollution.

It's a tough challenge, partly because many of the things that define a 'successful slum' in terms of its social mobility, are in opposition to the things that make them safe and affordable and capable of housing vast numbers of residents. For instance, it is only the fact that a bit of land is sewage-sodden and neglected that makes it affordable for a villager who has nothing but some ply sheeting to construct a shack. From this terrible one-room beginning, he has the space to live and set up a workshop or other business, and the possibility of eventually affording a concrete floor, hooking up to power cables, slowly replacing his plywood with breeze blocks, adding a room, renting out a space, growing his business, selling his house to someone poorer and moving somewhere better . . . The migrant who moves into an anonymous but secure high-rise has no access to the street to tout her business, cannot easily expand, and will struggle to get the connections to help herself improve on her arrival situation.

Around the world, the most successful migrant cities tend to be dense but not too high – streets of blocks around four to six storeys – with direct access to the street from buildings, which are clustered closely to schools, health centres, social services, parks and markets. They have excellent transport links to the economic and cultural core of the city, and they have expansion potential – the ability to add rooms or add a shop on the ground floor. Policies are also important: immigrants need to be able to start a business or be employed legally, and there needs to be an efficient, affordable system for their qualifications to be recognized so that trained surgeons aren't forced to work as cab drivers while patients are forced to wait months for medical procedures. Most important of all, there must be universal access to healthcare and education for all citizens regardless of origin and wealth, something supported by an abundance of research but which the United States still regards as a radical proposition.

With the number of migrants that will need to be housed, density is important and blocks several storeys high are the most efficient. Certainly, most Western cities could be built higher. The problem is that by increasing energy and space efficiency, we end up with a large deficit in opportunities for networking and business, which needs to be remedied. That means investing in social capital: ensuring a mix of large and small public spaces, including parks and squares, social clubs and local organizations, and policies to actively enhance inclusion. To ensure business potential, there needs to be affordable rent-or-purchase office, workshop or shop spaces. Importantly, it needs to be a mix of high- and low-rise, with business, retail, leisure and public spaces all integrated together. The failed experiment of concrete deserts of high-rise blocks that blight the suburbs of European capitals must not be repeated.

Migrants to the cities of the north should be welcomed in family groups. When immigrants arrive without their social networks, they can become isolated and potentially fall into dependency on criminal networks or prey to religious or political extremism. Families provide vital support and stability, broadening the networks that help people to establish roots in a new city; yet they are often refused entry under strict points-based immigration systems based on skills or wealth. Families also come with a diversity of skills, offering a range of potential benefits. Consider the skilled nurse who is sought after to fill a needed post and comes with her family: grandma can help with childcare while the mother works, grandpa is working in a restaurant passing on his skills, children are an investment in the future workforce, auntie and uncle work in childcare or as a gardener or cleaner, freeing other households to do their work. Each plays a role in oiling a larger economy.

Migrant cities are home to some of the most resilient, inventive and motivated populations in the world, but how that potential is harnessed and nurtured depends on government policy. Managed well, immigrant clusters will be a key driver of a city's regeneration and national growth during a time of climatic turmoil and global disruption; managed poorly, they will trigger social division and ethnic tension. To manage immigration means to invest in the housing, services and infrastructure required by the newly enlarged population. This takes away the pressure on resources and services that the native population would otherwise experience, and allows immigrant populations to live in dignity and contribute to the city's productivity. This investment itself may cause tensions in native populations who feel their facilities are being left out of the renewal and immigrants are receiving favoured treatment. The way around this is to plan upgrades to areas that include native and immigrant populations – everyone benefits from a new hospital or school – and to employ natives and immigrants in the building projects. In some new northern cities virtually everyone will be a migrant, either from elsewhere in the nation or in the world, which provides a perfect opportunity to build sustainable, socially integrated cities from the bottom up. Prefabricated buildings can be rapidly deployed, with many of their energy-generation and water-recycling systems built in, but municipal infrastructure will also need to be created and paid for to ensure we generate liveable cities rather than simply smarter slums.

This investment from government will be more than repaid by the successful integration of a vast new tax-paying labour force, and funds to assist city expansion could come from the new global body for this, the UN Organization for Global Migration (with powers), which would ease the pain. Painful or not, it is essential. Well managed, this vast migration would reduce poverty globally, protect millions from the worst effects of climate change, and create vibrant new cities from which to launch a good Anthropocene. The private sector can also help with costs. Canada already has a successful community sponsorship model, in which private or community organizations cover the financial costs and settlement support for humanitarian migrants. Canada has welcomed more than 300,000 refugees through its community sponsorship, and Australia is considering adopting the system in an overhaul of its dismal record on asylum. Seventy per cent of privately sponsored refugees in Canada declared employment earnings within their first year of arrival compared to just 40 per cent of government-assisted refugees. Migrants and host communities all do best when they are integrated into the labour force, and businesses have an important role to play. Some international companies are involved in global citizens' programmes, paying their employees to do voluntary work in a poor country; other companies also bring people over from poor countries as their employees. For instance Driscolls, a large Californian berry grower, employs a workforce in Mexico and the US, including a large number of Mexicans in its Californian plant, helping workers with training, housing, medical care and immigration issues.

Urban migration is recognized as the most effective route out of poverty – the largest study looking at this, conducted by the World Bank, concluded that for economic growth, there should be the highest possible urban population density and the growth of the largest cities through migration, with the study's caveat being that urban areas where rural migrants arrive must be given intensive investment and infrastructure development by governments.[1] The immigrant waves of the early twentieth century and post-war years coincided with a large expansion in public spending on education, health, housing, infrastructure, mass transportation and local government (and coincided with the decline in heavy industry that left neighbourhoods of low-cost housing). The gargantuan migration of this century demands similar investment.

There are places where governments have managed migration better than others. In the decade from 2000 to 2009, Spain's foreign-born population more than quadrupled, to almost 14 per cent of the total population, as the nation added 6 million immigrants. But unlike other European countries, Spain hasn't seen significant anti-migrant backlash despite its relatively high levels of unemployment and poverty[2] – most people believe that immigrants are needed (they make up one-fifth of the workforce) and that they are entitled to equal rights.[3] The reason is down to the way the government managed and prioritized its national immigration integration programme.[4] Spain believed in shaping migration policy in a comprehensive, planned way. This meant building genuine partnerships with other countries based on cooperation, not delegation; devising proactive rather than reactive policies; and leading public opinion rather than encouraging anti-migrant sentiment.

Take Parla, a sprawling low-rise city 20 kilometres south of Madrid, on the commuter route between the Spanish capital and Toledo. Once home to many of Spain's rural-to-urban migrants, it's now a city of international immigrants, many of them from Morocco and Latin America – some 4.5 million of whom arrived in the country during its economic boom in 2008. Unlike France, which made little effort to assist or manage its sudden influx of immigrants, or Germany, which largely ignored the needs of its migrants and blocked them from becoming citizens, the Spanish government actively invested in managing its mass immigration, with Europe's first policy initiative aimed specifically at making a functional, liveable migrant city. It began with citizenship: all fully employed migrants, included undocumented ('illegal') migrants, were made legal tax-paying residents with access to services. To deter dangerous illegal boat-crossings from North Africa, the government set up a programme in which tens of thousands of Africans were given Spanish work permits for one year. If their employment contracts were extended, they were then allowed to bring over their families and work towards full citizenship. It had an immediate and transformative effect, as Spain added half a million migrants to its economy every year, and these people were able to build lives, invest in a home, rent business spaces, send their children to school, and become active citizens keen to improve their lives and new home city, rather than an impoverished underclass, scraping a living illegally.

The Spanish government invested in a €2 billion programme to help the new migration work, including special education, immigrant reception and adjustment, employment assistance, programmes to help migrants find and build new homes, access to social services, healthcare, integration of women, community participation and community building. And the efforts paid off. The government's immigration spokesperson, Antonia Hernando, told the migration reporter Doug Saunders: 'These migrants are working legally now and paying the taxes that finance the pensions for a million Spanish people. They are the financial foundation of our country's welfare programmes, so we need to make sure that, in return, they have the same rights and livelihoods as other Spaniards.'

Infrastructure improvements and transport links ensured Parla's network could grow and its economy could function efficiently, with tramlines that run through the city and a high-speed rail connection to Madrid that takes just twenty minutes, opening Parla's economy to the greater city. The government has built large residential developments of mid-sized apartment buildings that provide ground-floor space for small businesses and pathways for home ownership. The mixed-use planning means the streets are busy, and the housing is of high enough density that there is good flow of customers to the businesses. But best of all, migrants feel that they belong. When the global economic recession hit, migrant-populated areas of Germany and France, already suffering high crime rates, erupted in violent protests. By contrast, Parla suffered no social unrest, despite very high levels of unemployment, because migrants there felt part of society. They had dignity.

Parla worked because the government anticipated a large influx of migrants and rather than trying to disrupt or control the migrant flows, it instead concentrated on managing this new population for the benefit of the national economy and society, and that involved investing in efforts to make it work. (In the past couple of years, Spain's immigration policy has become less positive, largely in response to the lack of support from its EU neighbours. However, surveys show that the Spanish remain positive about immigration.)

Inadequate or too-expensive housing is one of the factors that keeps people from moving to safer cities where they could be more productive, and much of that is down to planning and zoning laws that are not fit for expanding cities. Removing zoning restrictions would enable higher density housing and the mix of residential, business and public spaces that are part of the traditional attraction and evolution of a city. Higher population density is better for social cohesion and prosperity, as countless studies now show. However, getting permission to build within cities is difficult, because existing residents are often resistant and they pressure planners. This means housing gets built in more remote areas where opposition is weaker, instead of on well-connected sites where the need is greatest, which leads to car-reliant urban sprawl, reducing the productivity and other benefits of dense cities. In England, for instance, in every region, more land is now devoted to roads than housing.[5] Other restrictions, such as municipal codes across the United States that limit the number of 'unrelated occupants' who can live in a home,[6] also need to be removed, as they can prevent low-income people from sharing homes, splitting expenses and paying lower rents.

Relaxing business and licensing laws in cities in Europe and the US, so that retail, light industrial and commercial services can be mixed up with housing, would boost productivity and help people migrate successfully. The key factor for success in a migrant city is increasing the area's intensity – the amount of human activity allowed on a piece of land. Low-intensity neighbourhoods that zone off residential usage from other activities, particularly fashionable in the car-centric suburban sprawl of North America, can trap poor residents and immigrants in ghettos of hopeless poverty. By contrast, high-intensity neighbourhoods, from Hong Kong to Delhi, Manhattan and London, buzz with productivity and opportunity. Paris, with its rows of six-storey buildings, has a higher population density than New York City. London evolved out of a merging of individual villages, each its own walkable neighbourhood, whereas Manhattan's planned grid layout, a design which has its origins in ancient Chinese cities, uses discrete blocks of neighbourhoods – everything can be found within a short walk, as neither system zones off housing from retail or other activities. New migrant cities will need to integrate this intensity into their plans.

One place where this was learned late is Bijlmermeer, a planned city outside Amsterdam in the Netherlands. Known locally as the Bimmer, the original 1960s utopian vision centred on a vast honeycomb of thirty-one purely residential, massive tower blocks ringed by isolated parkland. The apartment blocks, with their confusing labyrinth of walkways, were devoid of public amenities, and even the roads were elevated, making access complicated and turning the ground level into a hostile desert. By the time construction was finished, nobody wanted to live there, and so it became a social housing 'dumping ground' for immigrants from Suriname and sub-Saharan Africa, many of whom lived on benefits with no obvious way of escaping a poverty trap. Bijlmermeer rapidly became known as the most dangerous neighbourhood in Europe, plagued by drug addiction, violent crime, murder and poverty. Then in 1992 an El Al cargo flight, doubling back to Schiphol Airport after engine failure, ploughed into two identical towers, killing forty-three residents. The disaster triggered campaigns to bulldoze its heinous buildings and redevelop the area.

Today the Bimmer is one of Amsterdam's most promising neighbourhoods. In place of the towers there are streets of close-packed mid-height apartments, each with its own garden, interspersed with spaces for shops and businesses. A new metro station and cycle routes now connect the area to the rest of the city. Cafés, subsidized theatres, art spaces and museums have sprung up, and the multicultural character of the area is promoted by the government, drawing people from all across Amsterdam to sample the food markets and restaurants. Early in the redevelopment, the government also invested in security for the area, and set up training schemes and business assistance, helping residents escape poverty with job opportunities and education. It worked: second-generation Surinamese migrants have rates of university education and income similar to those of native Dutch descent.

The inspired Chilean architect Alejandro Aravena fundamentally understood the needs of migrant housing when he designed his extraordinarily adaptable 'partial houses' in the port city of Iquique in 2003.[7] Commissioned to come up with an alternative to slum housing for rural migrants, his solution was to take a plot in the centre and build the essential infrastructure (which is so hard to retrofit into slums later), including sewerage, water and electricity connections. He then created basic concrete pads with all the core necessities – a roof, bathroom, kitchen – and space for residents to build on, bit by bit, as they could afford it. The partial houses were 25 per cent smaller than the average public housing unit in Chile, but with an extra-wide foundation, residents had plenty of room to expand. The government handed each family $7,500, which was just enough to buy Aravena's stripped-down models. As residents expanded their houses, their value grew. One study found that in its first two years, families had made an average of $750 in improvements per unit, doubling the size of their homes and raising the houses' value to an estimated $20,000 each. Although the project began with grey, ugly-looking identical concrete pads, within months, the neighbourhood had diversified with paint, additions and other improvements. Something like this could be a model for urban development in other migrant cities. In the Mexican state of Tabasco, for instance, construction company ICON has partnered with non-profit New Story to build affordable, earthquake-proof housing that can withstand extreme weather, using a giant 3D-printing machine.[8] The two-bedroom houses take just a day to be created by the machine, which uses a locally produced concrete, meaning a whole neighbourhood could be produced within a few months, with various print designs. Over time, families could then adapt and augment them to their requirements.

When 1 million refugees arrived at Germany's borders in 2015, Chancellor Angela Merkel had the option of sending the military to turn them away, or welcoming them into the country to help plug the national shortage of workers. Her famous response to the crisis was: '*Wir schaffen das*' ('We'll manage'). Germany needs at least 10 million extra working-age people for its economy in the next two decades. In a heroic effort, the majority Syrian refugees – many of whom were middle-class professionals who could afford to migrate – were resettled in Germany. It led to a brief surge in far-right politics, and there were initial mistakes, such as over where the migrants were placed. Instead of settling people in cities or neighbourhoods that already had large migrant populations, they were deliberately placed in empty neighbourhoods with available housing, including a barren ex-Communist high-rise block in a suburban outpost of Leipzig in the former East Germany, where there was no work and no hope of finding work. Happily, the proactive mayor of Neukölln, an immigrant-populated neighbourhood of Berlin, anticipated that many of the migrants would find their way to her district, and so she began preparing – schools were instructed to expect new children, for instance. And sure enough, the most enterprising of the émigrés soon arrived in Neukölln, where Syrian migrants in Germany have been a big success story – they have created more employment opportunities than natives. In 2021, when Afghanistan was overtaken by the Taliban, generating a wave of emigration, 62 per cent of Germans responded to the expectation of refugees with: '*Wir schaffen das.*' In March 2022, Germany was fast to welcome refugees from invaded Ukraine (regardless of their nationality), removing bureaucratic hurdles, providing free transport, and being proactive in accommodating their other needs. Germany is also changing its laws to allow immigrants to apply for citizenship far earlier – within three years of arrival in cases where there has been shown to be an 'exceptional degree of integration' – and to permit dual nationality.

Neighbourhoods need to be integrated into the wider community – this means ensuring that schools don't end up segregated through 'white flight', for instance by proactively creating excellent schools in the poorest neighbourhoods that attract all communities. But migrant residents also need agency to manage their own affairs, security, commerce and opportunity. Devolving power to communities is now recognized as a key catalyst for economic and social development – people already have a stake in their welfare; they need to be granted agency to become stakeholders in their cities.

Flexibility is key to making these city plans work. Urban migration is reducing the world's population growth, because people living in cities have fewer children than those living in rural villages. It is largely because of the speed of urbanization that global population is expected to peak in the 2060s, according to UN estimates – some experts believe it may peak as soon as 2050. This leaves us with an interesting challenge: how do we plan cities flexibly for growing, stabilizing and shrinking populations? At the same time, our demography is changing: globally we are moving from a younger to an older population, and older people have different housing and transport needs, for instance.

Tokyo is the world's largest megacity, but the emphasis for planners is on the hyper-local, rather than the traditional layout of a grand epicentre that radiates out to neighbourhoods of increasing insignificance and inequality. Local communities are involved in all aspects of the infrastructure and design of their neighbourhood, ensuring character and greenery through a process called *machizukuri*. With the demographic shift in Japan increasingly evident – it has more than 2 million people aged over ninety and adult nappies/diapers outsell infant ones – Tokyo is constructing 'daily activities areas', akin to school-catchment areas, catering to older adults. These are like villages within the city, where people can navigate all their amenities easily on foot. Tokyo manages to be a megacity that also operates at the scale of community interaction. In the UK, too, the shifting demographic is changing cities – retirement villages, long relegated to rural locations, are being brought into town centres, with planners banking on the 'grey pound' revitalizing high streets. New elderly-care developments are being built in city centres and vacant retail and office blocks are being repurposed, meaning the expanding population of over-65s will be in walking reach of shopping, leisure and entertainment. Cities across the north will need to cater to their ageing populations even as they expand to attract younger immigrants. Accommodating the elderly in the design and planning of our cities will ensure their sustainability for centuries to come – after all, the youthful migrants who help build them will themselves be enjoying them in old age.

9

Anthropocene Habitats

Migration this century will be to cities, and while social and economical sustainability is important, so too is environmental sustainability. We must make sure our cities are safe as the planet heats. And cities must also not make conditions worse: they currently consume two-thirds of the global energy supply and generate three-quarters of the world's greenhouse gas emissions. Some parts of some cities can be adapted for the new climate conditions; others will need to be abandoned or relocated; and new cities will need to be created to home billions of migrants.

Cities are particularly vulnerable to the impacts of climate change, experiencing the effects of heat, sea-level rise and extreme weather more acutely. Hard surfaces, such as concrete, absorb the sun's heat, and tall buildings diminish air circulation, while concentrated human activity (including vehicle engines, heating and cooling units) all add to the so-called urban heat island effect, in which cities experience hotter temperatures than surrounding areas. Urban temperatures are currently 1–2°C warmer than surrounding areas – and slum neighbourhoods can be at least triple this. The same hard surfaces of concrete and asphalt prevent rainwater being absorbed, so storms can quickly lead to flooding. Cities, of course, also concentrate more people per area than rural places, meaning more people are affected by heatwaves, air pollution and the devastation of extreme weather.

Of the 100 cities worldwide that are most vulnerable to climate change, ninety-nine are in Asia and eighty are in India or China. More than 400 large cities with a total population of 1.5 billion are at 'high' or 'extreme' risk because of a mix of life-shortening pollution, dwindling water supplies, deadly heatwaves, natural disasters and the climate emergency, according to a 2021 report by global risks consultancy Maplecroft.[1] As I've explained, the additional effects of humidity combined with temperatures just slightly higher than today will make equatorial latitudes intolerable.

Additionally, coastal cities that are home to around 60 per cent of the world's population are experiencing sea-level rise at a rate four times faster than elsewhere, because the sheer weight of their buildings and infrastructure pushes the ground down relative to the water.[2] Buildings and streets are sinking into the cavities created by construction work, causing flooding. Over the past sixty years, Shanghai (which means 'above the sea') has sunk by 2.6 metres, eastern Tokyo by 4.4 metres, Mexico City by almost 10 metres, and half of New Orleans, a city which is sinking at four times the rate at which the sea is rising, is already below sea level. These cities, which are currently absorbing migrants, will soon be generating them – and in large numbers.

Jakarta, the fastest-sinking city, is dropping at an alarming rate of 25 centimetres a year. The Indonesian government has decided on a mass migration as a solution. It will move its capital to a newly constructed city, to be named Nusantara, on higher ground on the forested island of Borneo. This multi-billion-dollar project aims to save Jakarta's citizens – who by 2050 will number 16 million – from the waves. However, the construction, which will take decades, will have huge environmental consequences for one of the most important planetary ecosystems, and still leave citizens vulnerable to extreme heat and fires.

Other cities are trying to hold back the waves with barriers and sea walls. Venice, built to accommodate the regular 45-centimetre rise and fall of tide in the Venetian lagoon, is now partially under water seventy-five times a year, with half of the major floods in its 150 years of record-keeping occurring since 2000. The government has constructed a barrier of submerged inflatable gates that can be raised during high tides to separate the lagoon from the sea. But the barrier has been designed to cope with no more than a 20-centimetre rise in sea level – something that could be exceeded as soon as 2050. Venice is already more of a museum than a living city – in the summer, around 60,000 tourists a day visit a city that's home to just 52,000 – as lack of investment and repeated inundations have led to a migratory exodus in recent decades. More than 120,000 residents have left Venice since the early 1950s, and over the last twenty years the pace has hastened. Soon it will be solely a museum. Other celebrated cities, or parts of them, will follow.

Cities have entrenched assets – there is enormous embedded wealth sunk into them, so there will always be financial imperative to shore them up even as residential areas submerge. Tokyo, Bangkok, even Dhaka and Lagos will not be completely abandoned. Instead they will become more engineered, with huge infrastructure investment. New York is planning the 'Big U', a vast sea wall to protect the financial district of lower Manhattan; but it would leave anyone living north of West 57th Street exposed to the waves. The city is already dealing with regular inundations, which in 2021 saw people swimming in flooded subway stations and geysers erupting out of the streets' drainage covers. One man described swimming out of a flooded station in Manhattan after a storm in 2005, while 'next to me fleeing was a bunch of rats'. It could potentially get far, far worse, with New York City facing the possibility of being a metre under water by the end of the century.[3] Rotterdam – already 2 metres below sea level – is planning another massive barrier system, along with floating houses. The densely populated atoll city Male, capital of the sinking Maldives, already has sea walls and other barriers. They protect the city – for now.

As King Canute demonstrated, the tide wins. The most vulnerable inhabitants of all these doomed cities are the poorest, including the migrant population – slum dwellers living in unsanitary housing and the rural people who join them, flocking to the city to seek safety when extreme weather hits. In other words, people today are migrating towards disaster. Cities have stronger infrastructure, more hospitals, and other essential services, so they are often seen as a refuge. Bangladesh's capital, Dhaka, is one of the most densely populated cities – about 40 per cent of the city's 14 million residents live in informal settlements, and 70 per cent of those were forced to leave their homes because of phenomena related to climate change, including cyclones and coastal and riverbank erosion. But Dhaka is itself no safe haven. As I picked my way through a still-sodden slum there, residents showed me how high the waters reach – eye level – flooding their homes and destroying their few possessions. The entire neighbourhood is forced to find refuge on the raised roads (which also flood since they lack drainage), sleeping exposed or in tents, during these inundations. And the lack of water and sanitation brings killer waterborne diseases.

When poor people migrate to cities they tend to get stuck there, having used all their resources in the move. While middle-class and wealthier people can afford to migrate to better locations, the poorest and most marginalized get trapped in the most vulnerable cities, unable to afford to move away. In 2018, the Migration Policy Institute conducted a comprehensive review of all the research evidence on climate and migration.[4] It found that climate shocks are highly likely to *reduce* a community's likelihood of moving (by hurting their ability to afford to migrate); when they do use migration as a survival strategy, it's almost always within the local region.

The solution is to plan: safer cities for expected immigrants, relocation strategies for risky districts, and ways to facilitate international migrations. Governments can help by withdrawing their backing for property insurance and buying back land. However, in many cases compensation and buyouts for individual households look woefully inadequate. Moving communities can take decades of planning to ensure the best transition for livelihoods.

One place that's taking it seriously is Kiribati, a state of low-lying coral atolls bisected by the equator, whose economy relies on fishing and coconut production. Over the past 5,000 years, these islands have been settled by waves of immigrants, from the early Austronesians to recent Europeans, who have built a rich culture. Now the entire population is preparing for mass emigration because of the dangerous sea-level rise. In 2014, President Anote Tong told me the country had reached 'the point of no return'.

Kiribati is pioneering the steps that multiple other cities and nations will have to take as they face the reality of unliveable conditions. It has purchased territory in Fiji for its beleaguered population and is also helping its citizens find new livelihoods in other countries. Tong began his 'migration with dignity' programme a decade ago, beginning to move people gradually though employment abroad, such as sending nurses to New Zealand. His aim, he explained, is to avoid turning his citizens into refugees with a large humanitarian evacuation during an extreme weather disaster, as has befallen other islands, such as Puerto Rico.

Tong talked to me about his 'duty and responsibility' to prepare citizens for the psychological – as well as practical – hurdle of leaving their ancestral land, graves and culture, including their familiar language, songs and stories. 'I am determined to help our nation adapt to what is coming, which means addressing risks, and strengthening our resilience to our islands no longer sustaining human life,' he said. 'We want our young people to be able to migrate voluntarily with dignity to other countries, so we're investing in the education and skills to equip them.'

Planning is key not just for new cities, or foreign migrants, but also to encourage migrants from other unsafe places in the nation to move to safer cities. In Louisiana, for instance, government officials are spending $48.3 million to relocate households from the low-lying Isle de Jean Charles to higher ground forty miles away, as part of the first federally funded, climate-change-induced community resettlement project in the US. New Zealand has a Managed Retreat and Climate Adaptation Act to help individuals and communities to relocate; the Canada-based Climate Migrants and Refugee Project is mapping displacement within and into British Columbia. In Bangladesh, too, government agencies are looking into creating migrant-friendly towns outside major cities, to reduce the pressure on places like Dhaka.

BUILDING RESILIENT CITIES

While Dhaka, New Orleans and Venice will become increasingly unviable, with residents emigrating elsewhere, plenty of other cities will be able to cope with the coming changes, and benefit from a newly mobile workforce by offering them a new home. To be clear, all cities will need to adapt even if they face relatively minor impact from climate change, because of the transition to net zero greenhouse gas emissions. Well-located cities will draw many millions of people with complex needs, requiring safe, sustainable homes. These cities will need to be environmentally resilient, ensuring tight efficiencies in resource use, minimal waste and the elimination of dangerous pollution.

The big risk for cities this century is extreme weather, and new developments must be appropriate to the risks. It makes no sense for migrants to escape drought in their homeland, only to move to a city that is vulnerable to flooding – it is simply trading one kind of climate risk for another.

High and low extremes in rainfall will become more frequent, and all cities will need to adapt so these events don't become catastrophes. Rain gardens, which capture and channel stormwater into underground cisterns or simple depressions in the ground, beneath planted rushes and other vegetation, have been installed in cities from New Orleans to London to cope with drought. The largest increases in heavy rainfall events are expected in high-latitude regions, including northern Europe and northern Asia. China's government has committed to 80 per cent of its cities having 'sponge' capabilities by 2030, at a cost of some $20 million per square kilometre. Cities such as Wuhan now use green spaces, marsh zones and underground storage tanks to absorb rainfall and prevent floods. Others construct canals, widen sewerage, install fast-flow drainage and use permeable paving and surfaces. Barcelona is re-landscaping swathes of road surface to better absorb rainwater and mitigate heat. Further north, Gothenburg in Sweden is embracing the increase in rain with new water-management infrastructure, as well as artificial waterfalls and a *Regnlekplatsen*, or 'rain playground', designed to be particularly fun when it's wet, including a way for children to make pools, rivers and dams. Other cities are meeting the rising waters with innovative floating infrastructure, including houses, hospitals and agricultural beds, which can rise and fall with the water levels. The Netherlands has several floating communities – the houses, often prefabricated, are fixed to the shore, often resting on steel poles, and are usually connected to the local sewer system and power grid. They are structurally similar to houses built on land, but instead of a basement, they have a concrete hull that acts as a counterweight, allowing them to remain stable in the water. The Maldives is also planning a floating complex off Male, including affordable housing for 20,000 people, designed by Dutch firm Waterstudio. Artificial reefs under each house will help support marine life, while air-conditioning units will use pumped water from the deep sea for coolant.[5] Flood-safe homes don't have to be expensively engineered rarities – the Bangladeshi architect Marina Tabassum has designed award-winning, flat-packed, raised housing for refugees – made from bamboo, yet storm- and flood-proof.[6]

Heat is another serious problem that cities will have to solve, ideally using 'passive' designs that enhance overall sustainability rather than adding carbon emissions. Demand for cooling will soar this century, becoming a key social justice issue, especially during heatwaves, when lack of access will prove deadly. Cooling already uses 20 per cent of global energy production, and this is expected to triple by 2050. Months of heatwave in the spring of 2022 across India and Pakistan meant hundreds of thousands of people were unable to work after 10 a.m., with load-shedding power outages leaving people without access to cooling or refrigeration. Cooling is not just going to be a problem in the tropics, where there is already fast-rising demand, but in today's temperate zones where vast populations will be headed.

Insulation will help manage this burden, and strategic use of water – used for cooling by architects and planners for centuries – will also play a role. Many cities are planning new canals and water features. In Omonia Square, the central plaza in Athens, analysis has shown temperatures dropping by up to four degrees since a multi-jet fountain was installed in 2020. Rooftop and vertical gardens provide a holistic solution to heat, biodiversity loss and extreme weather, and although they're most fecund in the tropics, using vegetation such as sedges works well in the far north. In Chicago, rooftop vegetation proliferated after new laws and incentives were brought in in 2004.[7] Half of the city hall is now covered with a roof garden and while summer temperatures on the uncovered area can reach 77°C, the gardened part stays closer to air temperature, about 32°C. Roof gardens can also capture rainfall, reducing stormwater run-off.

Painting roofs and other surfaces white also reduces heat. One study found that a clean white roof that reflects 80 per cent of sunlight will stay about 31°C cooler on a summer afternoon, and reduce indoor temperatures by up to 7°C.[8] A cool roof can save as much as 40 per cent of air-conditioning costs, the researchers calculated. Even in India, where most roofs are made from metal, asbestos and concrete, and temperatures can reach 50°C, lime-washing roofs managed to keep indoor temperatures up to 5°C lower. This low-cost tool could produce a cooling effect comparable to offsetting 24 gigatonnes of carbon dioxide – the equivalent of taking 300 million cars off the road for twenty years – if roofs were whitened worldwide.

White roofs will also play a key role in more northern cities as the world heats, and scientists are continuing to develop more reflective paints. The best so far reflects more than 98 per cent of sunlight. This is significant because every 1 per cent of reflectance on a roof works out as 10 watts per square metre less heat from the sun. So using ultra-white paint to cover a roof area of about 93 square metres, could produce a cooling power of 10 kilowatts, which is more powerful than the central air-conditioners used by most houses.[9]

These twenty-first century refuge cities will need not just to battle extreme conditions, but to do so while mitigating climate change. Buildings alone account for more than half a city's carbon emissions on average, and 70 per cent in major cities such as Paris, London and Los Angeles. By 2050, the goal is for all buildings to use only as much energy as they generate;[10] and the mayors of nineteen cities, including London, have agreed to get there by 2030.[11] It starts with insulation to avoid heat leaking through walls, floors and ceilings, as well as windows that reduce heat intake, and reflective roofs. For existing buildings, this can be time-consuming. The Dutch *Energiesprong* whole-house refurbishment is not cheap but wraps homes in insulated panels that snap on easily like Lego. Thermal wallpaper, which can be decorated over, is another option. Fully decarbonizing means replacing (and electrifying) inefficient heating and cooling systems, which are, in turn, responsible for more than half of a building's energy use, in addition to hot water and lighting. Heat pumps can be placed under parks, public squares, roads, rivers and canals in every city to heat and cool buildings. The city of Ithaca in New York has raised $100 million through an innovative investment programme to decarbonize all of its buildings while creating new jobs by 2030[12] – something more cities could try.

Zero-carbon new-builds are easier to design efficiently, and the rapid expansion of cities this century is an opportunity to innovate.

Melbourne's Pixel Building, which opened in 2011, has panels to control the amount of light coming into the building, and 'smart' windows allow heat to escape on summer nights while filtering in fresh air. Solar panels and wind turbines sit on the rooftop, generating renewable power. Canada's first carbon-neutral building, in Waterloo, Ontario, also has solar walls and a three-storey green wall to offset carbon emissions. Smart heat- and light-responsive materials and fittings will become standard in buildings, from outer skins that shade them during the hottest parts of the day and allow the sun in during cooler times, to floors that generate electricity from footfall, and rainwater systems that help minimize water loss.

New housing for expanding migrant cities is likely to be prefabricated and modular for ultimate ease of construction and flexibility of use and reuse, as the city's demography changes, especially if formerly uninhabitable cities become liveable again towards the end of the century. These prefabs can be made from organic materials, such as bamboo or fast-growing softwoods that are specially engineered to give them the strength and durability of harder materials. Building from wood actively locks in carbon, in contrast to concrete and steel, which are together responsible for 13 per cent of global emissions. One study found that using wood to construct a 120-metre skyscraper could reduce the building's carbon emissions by 75 per cent. Wood is also lighter, faster and versatile – wooden skyscrapers, or 'plyscrapers', are under construction across the globe from Norway to New Zealand, made from cross-laminated timber (CLT) stuck together with fire-resistant glue, which is as strong as structural steel and better at withstanding fires (steel can buckle and even melt). Most of our new housing could be kits of light, five- or six-storey blocks built from CLT, enabling rows of street housing to go up in days. The French government has ruled that all new public buildings must be made from at least 50 per cent timber. The Swedish town of Skellefteå has wooden schools, bridges, a skyscraper and hotels, even car parks.

Several companies already make prefabricated wooden housing complexes that can be delivered by lorry and rapidly assembled, such as IKEA spin-off BoKlok, which arrives with solar panels and other self-sustaining features. The beauty of trucked-in housing is that it can be trucked out again and redeployed where it's needed. In the dynamic housing landscape of mass migration, this could be very useful.

Government policy is essential to drive this transition, including through carbon-pricing incentives and the removal of fossil-fuel subsidies. Construction today is a heavily polluting industry; our new cities will need to be built with low-carbon cement-free concrete and steel manufactured using an electric arc furnace rather than combustion. 'Concretene', a graphene-enhanced concrete that is so strong that 30 per cent less material is required and no steel reinforcing, could also significantly reduce emissions.

Cities will have to work harder for many more people in less hospitable conditions. Water will need to be circulated, cleaned, stored and reused, just as it is in today's driest cities. Buildings will have to generate energy, prevent energy loss, act as scaffolding for climbing plants, homes for insects, birds and microorganisms, and protect their inhabitants from dangerous heat and storms. City living will mean dense apartments with private balconies, roof gardens or yards, and shared outside space. City landscaping will include water management and storage, with ponds and canals, and spaces for socializing. Transport will have to be largely by foot or pedal – including electric cargo bikes and rickshaw taxis. This is a particular challenge for North American cities, which tend to be low, sprawling and car-based. But with hundreds of millions of migrants needing new housing, this is an opportunity to build dense, transit-friendly communities. Where more powerful vehicles are required for longer journeys or heavier loads, they will have to be electric, and mostly pooled vehicles or rentals. Public transit will need to be cheap, frequent and electric.

Tragically, many cities in the tropical belt won't be able to adapt because conditions will be too extreme, and climate adaptation funds would be better spent in helping citizens adapt to a future elsewhere instead. As President Tong concluded, that means investing in education, so people find it easier to get employment in their new city; and governments negotiating land purchase or rental elsewhere.

Training is particularly important for rural migrants, who could easily end up begging on street corners or trapped in poverty, because they don't have the skills to thrive in the city. Bangladesh is already working on retraining programmes. While the older rural population is making climate adaptations, such as switching to farming salt-tolerant rice, or shrimp instead of vegetables, younger people are undergoing a 'second-order adaptation'. They are being given a tailored education to enable them to flourish in an urban environment – the government is preparing people to resettle in towns with greater protection.

Before long, cities in the demographically challenged north will be competing for migrant workers, and those that can offer employment, education and affordable housing will benefit. Migrants arriving with professional qualifications need to have an affordable way of enrolling in college to get their training recognized, but an engineer from Kabul, Afghanistan, who is working as a taxi driver in Duluth in the US can't afford expensive housing while he studies part-time at university to get his qualifications approved. Mid-sized or second-tier cities are primed to take advantage of the immigrant influx because they have cheaper housing and universities, and often have labour shortages. This is where many migrants will move, with cities, even in the far north, growing in size and importance as the diversity of their populations and knowledge base increases.[13]

Migrants will be attracted to jobs in growth industries that can be located in cities anywhere (unlike agriculture and mining, for instance), such as biotech and data management. Some of the newly identified 'climate haven' cities have already experienced unplanned immigration, as a result of the Covid pandemic. Nearly 11,000 people moved to Vermont in 2020, swelling the 624,000-strong population by around 1.5 per cent. This was a wake-up call for climate-migrant sceptics, who said: '"There are no jobs here, people don't have any motivation to come",' says Kate McCarthy, sustainable communities programme manager at the Vermont Natural Resources Council. 'With Covid, people have seen two things: you don't need to be coming to a job, you can be bringing a job with you; and we don't know what the next year is going to bring.'

In 2019, a coalition of ten cities – including Los Angeles, Bristol and Kampala – formed the Mayors Migration Council to help city leaders react to climate-driven urban migration at the local level, in ways that benefit both the municipalities and the newcomers. It means different things for different cities. The Canada-based Climate Migrants and

Refugee Project is currently mapping climate displacement within, and into, the state of British Columbia, so it can give cities concrete recommendations on how to prepare. For others, transformation means ensuring equity for migrants by redesigning housing and transport systems and ensuring a greater diversity of jobs – as is the case in Bangladesh. The industrial port city of Mongla is one of a number of second-tier Bangladeshi cities preparing themselves to receive rural climate migrants – some 4 million Bangladeshis were forced to flee their homes by extreme weather in 2020 alone. Climate migration offers cities like Mongla a chance for economic revival, and it's creating new educational facilities, housing and jobs, as the government tries to prevent worsening overcrowding in the slums of Dhaka.

Anchorage, Alaska, is creating a new set of migration policies, based on recognizing that climate migrants offer unique skillsets that come from surviving shocks and stresses. Most of the city's migrants come from the Philippines and other Asian countries, with 10 per cent originating in Mexico. Anchorage's first lady, Mara Kimmel, is an immigration attorney and believes migrants themselves carry a unique capacity for urban transformation that is hugely beneficial to the cities that welcome them. The city is boosting inclusion through language programmes, equitable access to transport that connects migrants with housing and work, and by matchmaking newcomers' skills with available jobs.

'The sustainability of these enlarged migrant cities is completely dependent on the speed and the effectiveness of integration of new populations,' says Neil Adger, professor of human geography at Exeter University in the UK, who has studied migrant cities globally, looking at how people grow to feel ownership of their cities. Migrants who are welcomed often become fiercely loyal citizens, helping strengthen the entire society. A 2019 US study found that immigrants and their children have levels of patriotism that are the same as or exceed those of native-born Americans, and they have more trust in American government. 'Immigrants bolster patriotism and national trust in American government institutions,' the researchers concluded.[14] This is borne out in countless examples, including the Irish-born French immigrant Samuel Beckett, who was awarded the Croix de Guerre for his heroism in the French Resistance, many of the UK's leaders – Prime Minister

Boris Johnson, Chancellor Rishi Sunak, Home Secretary Priti Patel and London Mayor Sadiq Khan are all first- or second-generation immigrants – and social reformers such as Thomas Paine, the British-born American immigrant whose pamphlets helped inspire the patriots to declare American independence in 1776.

Today, just one in seven of us is a migrant, and of these only 20 per cent crossed an international border. That number will soar over the next few decades as people concentrate in the world's viable cities. These citizens will depend on the unpopulated rest of the planet for everything else – most urgently, their food.

10

Food

One of the biggest challenges to mass migration is feeding people in their new homes. The UN calculates we'll need to produce as much as 80 per cent more food by 2050 to feed an extra 2 billion city dwellers.

But climate change effects and environmental degradation mean that many of the places where we currently farm will be off the table. And with agriculture currently responsible for around 15 per cent of global carbon emissions and accelerating biodiversity loss, we need to radically alter how we feed the world. We need to make the process of feeding ourselves more efficient and less environmentally destructive, and grow our food where it can flourish in a hotter world with unreliable water availability. This means adapting and improving food production in the global south, and establishing new and far bigger sources of food production for those migrating to the cities of the far north.

On average, each person needs to eat 2,350 kilocalories per day. Farmers worldwide grow enough food for everyone on the planet to consume 5,940 daily kilocalories. However, we waste a lot – 35 per cent of food. And one-third of the crops we produce is used to feed animals, which is a very inefficient use of land and calories. What is left and actually eaten by humans, works out at 2,530 kilocalories per person – this is still more than enough for everyone's needs, but these calories are, of course, not distributed fairly, and many people cannot afford (or don't choose) to eat a healthy diet. There are huge global disparities in agricultural production and in access to nutrition. North America produces eight times as many calories as its population needs to eat; sub-Saharan Africa produces about 1.5 times the calories it requires. Worldwide, some 850 million people go hungry, a number that's increasing, while more than twice as many are overweight or obese.

Today, humans harness more than a quarter of the entire biological productivity of the land – potentially heading for 50 per cent within decades. Over 80 per cent of the world's agricultural land is used for livestock, and one-third of all our abstracted water. It's devastating nature: today, 96 per cent (by weight) of all the mammals on earth are either humans or our livestock; just 3 per cent are wild animals. We've seen a crash in flying insect and bird populations over the past twenty-five years, almost solely attributable to agriculture, and rainforests are being cleared at a rate of 30 acres per minute.

Fish, the last wild animal that we hunt in large numbers, are also under huge pressure. Globally, 90 per cent of fish stocks are fully or over-exploited, with vast areas of the seabed resembling a desert, after subsidized fleets of bottom-trawlers scooped the life out of it. Artisanal fishers, many of whom fish sustainably, are increasingly pushed out of waters by these vast operations. Globally, we extract 80 million tonnes of fish from the oceans each year, and another 80 million tonnes are farmed. At the rates we are depleting fish stocks, eating wild fish will not be an option within decades. Unfortunately, the way we farm fish today is also deeply unsustainable. Fish are plied with antibiotics and fed with enormous amounts of wild fish or corn and soy.

This broken, unsustainable relationship between our environment and our food production is the culmination of a process that began with the invention of agriculture some 10,000 years ago, eventually enabling the vast population of today's Anthropocene. In the thirty years from 1820 to 1850, as human population passed 1 billion, it's estimated that 600,000 square kilometres of land in the Americas, Africa and Asia were opened up to farming – an area the size of Europe. Between 1850 and 2000, human population increased five-fold, something achieved through the so-called Green Revolution of high-yielding wheat and rice variants, chemical fertilizers, pumped irrigation systems and other modern agricultural techniques.

Today it takes just thirteen years to add another billion people to the world. Our triumph over nature would seem complete. However, we have pushed the planet out of the Holocene, which supported humanity's development of agriculture, into a new, hotter world of limited fresh water, unpredictable climate and a far greater population, where all the best land has already been taken. There remain limitations on how many people the planet's resources can feed even with the help of modern agricultural techniques. Earth's carrying capacity is currently set at around 9 billion, but in a world that is 4°C hotter, several scientists have warned that the limit may be just 1 billion, due to effects on crops, water supplies, extreme weather, sea-level rise and ocean acidification.

It's a sobering warning, and means we need to dramatically change how we feed ourselves.

Today, four-fifths of the planet's ice-free land is used to grow our food. Of the 300,000 species of edible plant, we rely on just seventeen species to make up 90 per cent of our diet. Much of this is farmed monocultures of cereals, produced by depleting aquifers, exhausting soils, killing pollinators and other insects, and polluting waterways. The essential activity of creating our food also involves the inhumane treatment of animals, and is mired in social desperation and poverty that leads to farmers taking their own lives.

All of these troubles pale in comparison to the impacts of climate change on today's food-producing areas over the coming decades. A recent study found that climate change has been holding back food growth for decades[1] – the world has lost 21 per cent of food output growth, equivalent to seven years of productivity growth, over the past sixty years.[2] The percentage of the planet affected by drought has more than doubled in the last forty years – affecting more people than any other natural hazard, most of them farmers. Globally, drought now severely affects crop production on every continent – including 80 per cent of agricultural land in the US. To date, this has been alleviated mostly through pumping up water from aquifers, which are running dry. This has far-reaching consequences. India has vastly expanded crop irrigation over the past decades, largely through pumping groundwater, which has resulted in regional climatic changes as the water evaporates and rains down elsewhere. Around 40 per cent of the rainfall in East Africa is now attributable to India's unsustainable groundwater extraction – a bounty that has enabled Ethiopian farmers to expand crop production to new areas. With aquifers drying up completely in India over the next five to twenty years, the catastrophic costs will be felt by farmers in East Africa too.

One-third of global food production is threatened by climate change this century, according to an analysis in 2021.[3] At a global temperature rise of 2°C, an additional 189 million people will go hungry; at a 4°C rise, the effect is ten times worse with an additional 1.8 *billion* people in hunger. One study finds that every 1°C increase in temperature will lead to corn (maize) crop losses of 10 per cent in the US alone, in addition to the drops in global wheat, soya and rice yields.[4] Some researchers think this is a vast underestimate, because increased insect attacks could push per-degree losses up to 25 per cent. Some 20 per cent of the world's land was affected by locust plagues in 2020 alone, affecting twenty-five countries across the Horn of Africa, the Arabian peninsula and into the Indian subcontinent, a huge region where 24 million people are food insecure and 8 million are internally displaced.

The oceanic buffet will also be hit. Marine heatwaves are already altering ecosystems, bringing tropical fish into temperate kelp forests and destroying coral-reef fish nurseries. At 4°C, the frequency of these heatwaves increases forty-fold, and they last for a third of a year on average, and cover an area twenty-one times larger than today. One study projects that in a 4°C-hotter world, many tropical marine eco-regions will become dead zones, since ocean temperatures will be above the thermal tolerance threshold of all species.[5] The highly productive Southern Ocean encircling Antarctica will be one of the regions most and earliest affected, with 90 per cent of its area becoming too acidic to support shell-building organisms, including coral and many species of phytoplankton, which are the basis of the marine food chain. Factor in the other effects of hotter, more carbon-rich seas, including jellyfish explosions and toxic algae blooms, and things become even more serious: the oceans will lose their ability to draw down carbon from the atmosphere, and the essential turnover of surface and deep-water layers, which circulates nutrients and stores carbon, will be lost. Biodiversity in the oceans will crash, and that includes our fisheries.

A GREAT FOOD TRANSFORMATION
IS NEEDED

Given the constraints over land, food and population over the next century, we need to dramatically slash waste – even cutting it by half would add 20 per cent more food to the world's supply. This can be done in the global south by investing in better infrastructure, such as improved roads to cut journey times, and more efficient technologies, refrigerated storage and sealed, dry containers. I'll never forget the distressing absurdity when I was in Uganda, of malnourished, hungry villagers in the drought-stricken north of the country, while unharvested fruit and vegetables rotted in the south of the country – all for want of decent roads to connect the two in a timely manner. Farmers in the south didn't want to risk paying transportation for produce that would rot before it got to market. Meanwhile USAID was flying in relief packages to northern villagers on the brink of famine.

There are solutions. Researchers are developing a way of using renewable energy to compress air into a liquid for use in affordable refrigerant units (and for air conditioning), slashing the waste of perishable foods. Many farmers don't have access to barns to keep their harvests dry, so toxic mould grows on the damp grain. Building a few million barns, 300 mid-sized storehouses and 100 large warehouses would cost around $4 billion, but slash food losses across sub-Saharan Africa by 40 per cent, according to one estimate.

In the rich world, we need a cultural shift to buying only what we will eat. Part of the problem is that food has become so cheap we don't value it. Denmark managed to cut food waste by 25 per cent in five years between 2010 and 2015 through a concerted effort of campaigns, fixing supermarket bulk discounts on perishable foods, encouraging restaurant-goers to take leftovers home in a 'doggy bag', and reforming best-before dates. The nation is now aiming for further cuts of 25 per cent by 2030. Rather than sending food waste to landfill or composting, a better option would be to feed it to insect larvae, such as fly maggots, which can then be fed to fish in fish farms or, better still, directly to us. Maggots are 40 per cent protein and 30 per cent fats, making them ideal to use in place of other animal products in processed foods such as burgers, cakes and ice cream.

With limited agricultural land available, by far the most effective and biggest change we will make will be to adopt a plant-based diet in which meat and dairy are expensive luxuries. This will immediately free up 75 per cent of today's agricultural land and slash carbon emissions and nitrogen pollution. In wealthy nations, farming is the most polluting industry – more so than oil companies: agriculture accounts for 0.7 per cent of GDP but 11 per cent of carbon emissions. Just by replacing some of the meat in our diets we could cut 70 per cent of greenhouse gas emissions associated with food production. One of the ways to hasten this would be to carbon-price meat, as we do coal. We don't need to stop eating meat completely – livestock will still play a role in agriculture – but farmed animals will need to be far fewer, free-ranging, grass- and pasture-fed, with a little seaweed added to their diet to reduce methane burps. Wild foods such as fish, and cattle products such as dairy, will be priced based on their availability and environmental impact, and thus, in most places for most people, rarely eaten – as caviar or game birds are today. After all, we will lack the grazing areas or resources to keep farmed livestock at anywhere approaching today's scale.

However, this will not be a hardship or deprivation because our nutritional needs can be met fully and easily without animal products, and our palates amply placated with a variety of alternatives, all of which have far lower environmental costs.

There are signs the transition has started. There are already a wide variety of meat substitutes, particularly for processed foods, made from plant and fungi proteins such as nuts, soya and pea. While soya likes a long, warm growing season, and has struggled to grow in northern regions, conditions in most of northern Europe and Canada, for instance, will be able to support the crop within a couple of decades. Peas, meanwhile, are already tolerant of temperatures as low as −2°C, so agricultural expansion for these meat substitutes shouldn't be restricted by our northern migration. A wide variety of plant-based dairy products is also disrupting the global livestock market, which is worth $1.2 trillion-a-year but receives $100 billion per year of meat and dairy subsidies. Once they dry up, investment in alternatives will boom. Pat Brown, CEO of Impossible Foods, is ambitiously targeting 2035 for the end of industrial meat farming and deep-sea fishing. The alternative, as the World Resources Institute calculates, is a world in which we would need 600 million more hectares of cropland and pasture – an area bigger than the EU – by 2050 to enable continued meat and dairy consumption. In other words, there is no alternative.

Producers are using biotechnology to create fake meats that bleed like beef – the Impossible Burger is made from a soy protein with a yeast that has been genetically modified to produce leghaemoglobin, an iron-carrying molecule like haemoglobin that gives the burger its meaty bloodiness. However, most of what we enjoy about meat is the taste and aroma of the Maillard chemical reaction: this is the fusion of sugars and amino acids that occurs when the food browns during cooking. This can now be convincingly replicated with plant-based molecules. For those who want to bite into flesh, the next generation of lab-grown meats will reach mass market later this decade, as huge investment in this new industry grows – it increased sixfold in 2020 alone.[6] The meat is created from single self-dividing muscle and fat cells that are grown in chains that are layered up, stretched and relaxed in a growth medium, until a steak accumulates. The advantage of this is that the labs can be located anywhere, and the biotech industry will be a large job creator for migrants in many new cities. Investors, including Google co-funder Sergey Brin, are betting on huge bioreactors to produce a cheap range of popular meat cuts for a fraction of the ecological cost. Some 80 per cent of people in the UK and US are open to eating meat produced in a factory rather than a field, according to a study in 2021, with the researchers concluding that cultivated meat is likely to be widely accepted by the general public.[7] Lab meat will likely be a luxury product, though, because of the hefty energy costs involved.

The United States could reach peak meat as soon as 2025 and, according to a recent report by researchers for the Boston Consultancy Group, within fifteen years the rise of cell-based meat will bankrupt the US's beef industry, at the same time removing the need to grow soya and maize for feed. By 2035, the report predicts, an area a quarter of the size of the continental US 'will be freed for other uses' as livestock farming colapses.[8] In a hotter world where so many regions will be impossible to farm crops, such efficiencies will be essential to feed our growing population.

Fish farming will continue to be important in the coming decades, but we will have to reform today's extremely problematic methods. Open-pen salmon farms consume a lot of wild fish in feed, produce a lot of waste, suffer from huge infestations of fish lice, and the fish escape into the wild, contaminating natural ecosystems. There are now more farmed than wild salmon in the Atlantic. Land-based fish-farming could solve these problems, using a recirculating aquaculture system, with temperature-controlled water pumped in and out of the fish tanks, and insect-based feed. These new systems, which are currently being built in the Maine cities of Belfast and Bucksport, could help revitalize some expanding northern migrant cities. A land-based fish farm can be housed almost anywhere in a few multi-storey buildings, with the relatively high energy costs of its system met by renewables. This sort of farmed fish will be expensive, but as we've seen, fish and animal products will be occasional meals rather than part of our daily diet in the more sustainable decades to come.

The lowest-impact meat comes from insects, which are currently enjoyed by 2 billion people in 130 countries. If you eat anything with carmine, a red food colouring in sausages, pastries, yoghurts and juices, it's likely you've been eating the cochineal scale insect. The bugs are farmed on cacti in Peru in an industry worth some $38 million per year, which supports more than 32,000 farmers. Insect farming has huge potential as a sustainable animal feed and an addition to human diet, and creates useful by-products that can be used as fertilizer and material for medical purposes. Insects can be bred in significant numbers without taking up large amounts of land, water or feed – in fact they can be fed on our waste products, including human sewage, in a pleasing example of the closed-loop economy.

Consider that the average farm animal converts just 10 per cent of the calories it eats into meat and dairy foods, and just 25 per cent of the protein. Crickets or black soldier flies, by contrast, need six times less feed than cattle, four times less than sheep, and half that of pigs and broiler chickens to produce the same amount of protein. Insects produce body mass at an astonishing rate, in part because as cold-blooded animals they don't need to expend energy on regulating their body temperature. In particular, insect farming has huge potential for farmed fish feed, providing a higher-quality, protein-rich substitute for existing, unsustainable wild-catch fish protein. And insects make a much more efficient livestock feed than grain: it takes about a hectare of land to produce a ton of soy per year; the same area could produce up to 150 tons of insect protein. The industry has attracted significant investors in the past five years, hoping to disrupt the $400 billion global animal-feed market.

As migrant populations concentrate in cities in the north, insects will be the most versatile and appropriate livestock. Black soldier fly larvae could be farmed in multi-storey buildings or basements located around our cities, where they can make use of municipal waste streams. The entire insect is edible and the powder produced is a superfood, high in protein and essential fats, and rich in micronutrients such as iron and vitamins. This will be a primary source of protein and fats for our 9 billion population by mid-century.

Most people will transition to a plant-based diet over the next decade with little effort or conscious decision-making on their part, given the right nudges. One study showed that where menus are 75 per cent vegetarian, meat-eaters will tend to order plant-based dishes. I am not a vegetarian, yet I mainly eat plant-based foods, use plant-based oil and butter substitutes, and make my porridge with oat milk. The main occasions I eat meat or dairy are when I go out to eat, and choose something from a menu that I wouldn't cook for myself – it's a treat. The point is, I haven't made any difficult decisions to eat a mainly plant-based diet. Neither will you.

Most diets will become plant-, fungus- and algae-based, because that's the most efficient way to produce food for 9 billion people. With farmland under so much pressure because of drought, sea-level rise, extreme weather and temperatures too hot for labourers to work the land, we will need to adapt from Holocene to Anthropocene food production. For instance, we need to source food from the oceans. By that, I don't mean overfishing already exhausted stocks: mussels can be farmed in waters off coastal cities, for example, helping to clean the water. And photosynthesizing marine plants and algae are some of the most sustainable foods available and poised to become huge.

Seagrasses, which are the only flowering plants to grow in ocean water, have edible seeds that have been enjoyed by indigenous people for centuries, but only recently discovered by European chefs. The grain is highly nutritious, gluten-free, rich in omega-6 and omega-9 fatty acids, contains 50 per cent more protein than rice per grain, and grows without fresh water or fertilizer. Farming this 'marine rice' could expand agriculture in places where rising sea levels have made traditional crop farming impossible, like Bangladesh – some of the southeast Asian seagrass grains are as big as nuts. And seagrass has additional benefits in helping prevent coastal erosion, as an important habitat for marine biodiversity, and in carbon capture – it stores carbon thirty-five times faster than tropical rainforests and absorbs 10 per cent of the ocean's carbon annually despite covering just 0.2 per cent of the seabed.

Algae, whether seaweeds harvested from the oceans or micro-algae (also called phytoplankton) such as spirulina grown in industrial vats, are highly nutritious and have twice as much protein as meat. They are extremely fast-growing, absorbing carbon dioxide, but unlike other farmed foods they don't take up precious land. Kelp forest plantations, for food and biofuels, are already under way from California to the UK, some tended by submarine drones. These could be expanded along northern coastlines, providing a new industry for migrants as well as valuable food. Micro-algae growing vats can be placed pretty much anywhere on the planet, including deserts or underground – and algae can be dried and added to all kinds of foods, from bread to smoothies. They can also be used to tackle malnutrition or fed to animals, including farmed fish. Specially engineered bacteria can also be grown in vats to produce meat-identical proteins and fats efficiently and on a tiny footprint, using few resources. Some of these hydrogen-oxidizing bacteria have even been designed to extract the water and carbon dioxide they need from the air (without requiring the sun for photosynthesis).

We need to look beyond the Holocene technique of chopping down a forest to create bare earth, sprinkling seeds and letting the sun and rain do much of the magic. Cultivation of algal mats, which grow on the surface of ponds and lakes, and crops grown on floating platforms and in marshland will help feed cities. Cities will also need to contribute more to food production, with roof vegetable plots and vertical farms, which have the added benefit of cooling and cleaning the air. Deserts systems can play their part, with enclosed, self-sufficient greenhouse systems that circulate air and water, powered by sunlight. These are already used in places such as Australia and Jordan, and could help feed communities in habitable zones, such as northern China, who would otherwise have to migrate due to the desertification of their farmland. New materials, such as graphene, and more efficient desalination techniques can help optimize solar-powered closed-cycle agriculture.

Much of our agriculture will need to be at higher latitudes in places such as Canada or Patagonia because the tropics will be too hot for farmworkers. Most polar land has thin, rocky soils with little prospect of yielding much harvest; however, one study finds that three-quarters of the current-day boreal region will be suitable for growing crops in a 4°C-hotter world.[9] Agriculture would shift into Arctic Canada, Alaska, Siberia and Scandinavia, with cultivable areas moving as much as 1,200 kilometres north of current croplands, generating 15 million square kilometres of farm-suitable land (equivalent to that in the EU and US combined). Destroying the existing boreal forest and ploughing permafrost and tundra to sow cereals would be far from ideal. Instead, most northern agriculture could be focused on Canada's western prairies and the expansion of existing agricultural areas in Nordic countries and Russia, especially in the more temperate areas close to the Arctic Ocean.

This has the potential to shift global geopolitics. Many of today's biggest producers, such as the US and Brazil will see their productivity slashed. Meanwhile Russia, which is already the world's biggest exporter of wheat, will see its agricultural dominance grow as its climate improves.

Moving agriculture further north does mean dealing with less intense sunlight, however, particularly in winter. Since these are the same places we're putting our people, it also means that crops will be competing with cities for the same land and water. However, plenty of studies have shown that crops thrive with artificial light delivered by LEDs at exactly the right frequencies for photosynthesis. This means we could grow crops such as vegetables through the winter months, if need be, hydroponically in smaller spaces – stacked up in warehouses or even underground with renewably powered lighting – leaving valuable land surfaces for other uses. Indoor industrial systems using genetically engineered microbes and chemical feedstocks will help supply proteins, fats and other essential nutrients for our larger population, with land-based crops giving additional texture and flavour.

In areas of the world, such as India or Thailand, where dangerous wet-bulb temperatures will eventually make it impossible for farmers to live and work, agriculture could remain viable (where there is sufficient water supply), with remotely controlled robotic farmers, drone seed dispersal and AI-directed machinery for harvesting, maintenance and production. Colorado is already trialling livestock farming by drone.[10] For parts of the world where economies are still largely agricultural, the idea that farmers will no longer be able to work the land in person is cataclysmic: billions of people's lives and livelihoods – their sense of identity – is tied to the land. And the implications for food supply are terrifying. As we continue to heat the planet with our still-rising carbon emissions, the need to adequately plan for the migration of people and food production could not be more urgent. It means that in locations where farming is possible, efficiency is vital: every acre matters.

Modern agriculture, for all its sins, has dramatically improved yields – we'd need 2.5 times as much cropland to produce the same global food volume if we were using the methods of sixty years ago. Food production this century needs to be more intensive and industrial – but not in a way that over-uses fertilizers and water. That means closing the yield gap, which in sub-Saharan Africa can be as high as 81 per cent – Ghanaian corn crops have the potential yield of 8 tonnes per hectare, but actual yields are just 1.5 tonnes per hectare. Even in the US, the yield gap can be 40–50 per cent. To date, most attempts to close the gap have focused on even heavier use of water, chemical fertilizers, pesticides and fungicides on monocultures. However, while this is important in places where fertilizers are underused, such as sub-Saharan Africa, overuse is very damaging to ecosystems and its yield benefits don't last long before the soils become exhausted. This is partly because soil is full of microbes that improve plant growth in a number of surprising ways, including by enabling plants to communicate with each other and to receive nutrients they require at the optimal time. Ploughing intensively and using antimicrobial chemicals destroys this valuable soil ecosystem, reducing yields.

Farming will have to be intensive, but smart. For instance, pre-treating seeds with microbes before planting can help increase yields, particularly in drought conditions. Scientists are also developing new crops, such as perennial cereals that don't need to be dug up and reseeded each year, thereby helping keep soils intact, protecting their fertility and reducing carbon emissions. Planting wild flowers, such as clover, alongside crops increases the health of pollinating insects, and so the crop yields too.

Growing crops in this new, hotter world will mean selecting for varieties that are heat, drought and salt-resistant, through breeding and genetic modification. Creating crops with roots that are capable of fixing their own nitrogen like legumes can, for instance, would mean far less fertilizer use. Genetic research will help produce food with lower greenhouse gas emissions and less water use. It could also one day enable farmers to create rice and other cereals that are far more efficient photosynthesizers, so more can be grown on the same land. If scientists manage to make versions of our main cereal crops that are as efficient at photosynthesizing as corn and sugarcane, we could dramatically increase food production. It took us thousands of years of breeding knowledge, trial and error, and expertise to generate the Holocene crops we rely on today. Now we must generate new crops for our hot-house world that can feed billions more – and we have just decades to do so.

Heat-tolerant and drought-resistant crops, such as varieties of cassava and millet, will replace many of the current staples such as unmodified rice and wheat in our diets. Higher carbon-dioxide levels will mean they grow faster and need less water. We will need to diversify crops and use rotations to maintain soil health and fertility. Far more resources need to go into finding and storing different varieties – a crop-disease pandemic could be just as devastating for us as Covid. Crops will need to be selected for the environmental constraints we're facing, bringing an end to anomalies like unsuitably thirsty crops, such as cotton, being planted on water-scarce land.

Rice is a case in point. Flooded paddy fields currently account for about 6 per cent of all greenhouse gas emissions from the food chain – more than twice those of any other cereal – because so much methane is released from the flooded soils. Methane is roughly thirty times more potent than carbon dioxide as a greenhouse gas. On current population growth trends, emissions from rice growing could increase 30 per cent or more in the next twenty years. However, there is at least one variety of rice that could be grown without flooding, with a process known as SRI (system of rice intensification) cultivation. This uses fewer seeds and fertilizers, and needs less water. In a three-year study of 50,000 farmers using SRI methods in thirteen West African countries, farmers saved up to 80 per cent of the cost of seed and got an average 70 per cent increased yields and 41 per cent increase in incomes, while reducing methane emissions by 50 per cent.

In the UK, which is projected to experience more extreme weather events, including monsoon-type downpours and flooding, so-called wet-farming is being introduced. Rather than draining peatlands, which releases vast stores of carbon, crops are being selected that thrive in saturated soil, encouraging the re-wetting of peatlands. These include bulrush and reeds, but also sweet manna grass – similar to wild rice – which can be milled for porridge.

Wherever our food is produced, we will need to use far more precise nutrient and drip-irrigation systems to avoid polluting ecosystems, and also to reduce food loss and waste. This means using cover crops, mulches and intercropping, which recycle nutrients so that chemical assistance can be used sparingly and appropriately when needed. It means rehabilitating exhausted farmland and enabling unsuitable land to rewild. China helped improve its farmlands through a vast programme of integrated soil-system management that ran from 2005 to 2015, involving some 20 million farmers across 40 million hectares of land. The result was an average increase in crop yields of more than 10 per cent, while nitrogen fertilizer use dropped by 16 per cent, culminating in an economic saving of $12.2 billion.

Food is not just essential for life, it is central to our lives. Farmers across the world are undergoing a transition from a livelihood that sustained almost all of us for around 8,000 years to an urban existence where they are fully dependent on strangers to feed them. As a result, the most common way to protect against income poverty is by holding on to land – but this makes migration harder. In many countries, families hold ever-decreasing, sub-divided plots of land long after most of them have moved to the city, as an insurance policy (often because they can't afford urban housing). As a result, high-yield farming on large pieces of land is impossible. Rural livelihoods become harder and the cycle continues. At the other extreme, in rich countries a few wealthy landowners control vast areas of the nation, often not producing food from them, but making it harder for others to make a living in the area.

Like other kinds of wealth, land ownership is becoming concentrated into fewer and fewer hands. One per cent of the world's farms control 70 per cent of the world's farmlands, and are integrated into the corporate food system with little connection to the land, which drives destructive practices purely for short-term profit.[11] Tech giant Bill Gates is the biggest private owner of farmland in the US. Indigenous land managers, by contrast, prioritize protecting the land for the next generation while meeting the needs of the present – these land managers sustain 80 per cent of the world's biodiversity.[12] One solution to the unequal distribution of land is a land wealth tax, which would prompt landowners either to sell their land for useful funds or lease it to others who would make better use of it. Additionally, charging and taxing land owners for water extraction and for environmental costs such as discharging into rivers or emitting nitrates and greenhouse gases would help clean up agriculture. As an incentive, environmental services, such as maintaining waterways, planting biodiverse systems and protecting rare species should be remunerated.

As it becomes impossible to farm, rural farmers need to be assisted to migrate to the city through social security policies that provide a basic income. India, for instance, supports people through the Mahatma Gandhi National Rural Employment Guarantee Act (MGNREGA), which guarantees people the 'right to work', with 100 days' work at minimum wage. We will still need agricultural labour, of course, and many international migrants will have useful farming skills, but they won't necessarily be applicable to the conditions and agriculture of their new home. It is therefore essential to plan training programmes alongside settlement schemes. You can't expect a Bangladeshi rice grower to become a Scottish kelp farmer overnight, but millions of hungry people will depend on the success of rapidly transferred skills.

Many of the vast number of migrants to the northern latitudes will find work either in the burgeoning biotech industry or in the new agricultural industries – assisting in underground algae farms, or high-rise, climate-controlled, indoor vertical farms, or out in the countryside working the soils as all of our ancestors did. The business of making food has always been the most important human job – we've certainly made it harder for ourselves, but we do have the knowledge and technological expertise to innovate new ways of feeding ourselves through the coming crises. The question is whether we will manage the transition through calm preparation or wait until hunger and conflict erupt – an unconscionable outcome that would endanger us all.

11

Power, Water, Stuff

When large parts of the globe become uninhabitable, and bigger populations are concentrated more densely in strategic areas, the world's supplies of fresh water, raw materials and power will have to be more geographically focused. We will also have to be more judicious and sparing with our resources, keeping everything circulating for longer through recycling and reuse. Socially, we also face huge challenges, and this will need to be addressed through more equitable sharing of wealth and resources.

Let's begin with energy. Today, our technologically reliant societies use nearly 600 exajoules (around 25,000 terawatt hours) of primary energy globally (expected to rise to 39,000 terawatt hours by 2050). This has helped make us healthier, longer-lived and more productive. However, there are two big problems. First of all, the power sharing isn't equitable: hundreds of millions of people lack access to reliable energy – they haven't got electricity for lighting, cooling, or powering computers or refrigerators, nor a safe way of cooking and keeping themselves warm. People without access to sufficient energy are trapped in a life of poverty and ill-health, with devastating effects on their local environment. One of the biggest motives for deforestation, for instance, is to obtain firewood for heating and cooking, which is also a leading source of air pollution.

Many nations in the global south have discovered new domestic sources of energy in the past couple of decades, including oil reserves, rivers with hydropower potential, and solar and wind opportunities. The problem is that significant capital investment is required to realize this potential; grid infrastructure is extremely poor, so those most in need of power are the least likely to receive it; and many of these energy sources come with problematic environmental consequences. Mass migration, though, could well speed up energy access as people move to cities where they are easier to reach.

The other big global energy problem is that it is responsible for 87 per cent of greenhouse gas emissions, with all their catastrophic consequences. One in three of the carbon dioxide molecules in the atmosphere was put there by us. In the past fifteen years, we have added a third of the carbon dioxide to the atmosphere that humanity ever produced. By 2035 we are likely to have exceeded 1.5°C of average temperature rise. Every rise in global temperature pushes more people to migrate. Yet, by 2100 we are expected to use seven times more energy than today – partly because as poor rural people migrate to cities, they use more energy.

Decarbonizing the world's energy is the work of the next two to three decades. By the time my kids are in their thirties, this energy problem should have been solved. The problem is at once overwhelmingly enormous, but also solvable within the lifetime of most mortgages.

The first step is to decarbonize electricity production; the next is to power everything possible with electricity. Meanwhile, we need to capture any greenhouse gases created in the production of energy.

Even without the upheaval of human population movement, adapting cities to a net zero carbon economy during a time of global heating and extreme weather is an enormous undertaking. By the IEA's calculations, to reach net zero by 2050 we have to deploy new renewable capacity at a rate two to three times higher than the 'exceptional' 40 per cent growth rate achieved in 2020.

For poor countries with insufficient power capacity that are reliant on coal, the capital costs of deploying new renewables – even at their cheaper prices – can still be prohibitive. This is something that the international community should help with – by making cheaper finance available, for instance. The manufacture, construction, deployment, recycling and maintenance of renewable power technology will be a leading job creator in existing cities, but also in the new economies of the far north.

Our net zero world will be much more heavily reliant on electricity production, to power not just homes and businesses but also industry and transport, which currently rely on fossil fuels. The energy to make all this electricity can come from vast arrays of solar and wind-power plants installed in a belt across the uninhabitable desert regions of the mid latitudes. High-voltage direct-current transmission lines will then relay this power to the cities in higher latitudes, as well as to regional cities. Models for such systems already exist – Australia, for instance, is currently building the world's largest solar battery installation in its northern desert, which will send 24-hour renewable electricity to Singapore via a 4,500-kilometre undersea cable system, to be completed by 2027. Australia already produces far more solar power than it can use or store. North Africa is also sending high-voltage direct current to Europe from its solar arrays, and more regional transmission is promised from hydropower plants to cities. Africa's deserts are prime for large-scale solar and wind power – Morocco already has the world's biggest concentrated solar power project, in Ouarzazate, with others under construction. Solar and wind power require far less human interaction than fossil fuel power stations, so large areas that have become uninhabitable can still be used to produce energy for people living in safer locations thousands of kilometres away. Maintenance can be carried out by automated systems and robots. Offshore wind farms in the North Sea and Atlantic will also be connected into large regional networks that help balance supplies from other sources. Greenland could supply geothermal, hydro and wind power via subsea cables to Canadian, Nordic and British cities, for instance.

Tapping the immense power of the oceans would create another useful energy source for the ample coastline of the north. The EU expects around 10 per cent of its electricity to come from wave and tidal by 2050; if the systems work well and can be easily deployed, this could rise significantly.

There is also huge scope for solar panels on the roofs of buildings, vehicles and other infrastructure to be interconnected in distributed networks, turning neighbourhoods into virtual power stations to recharge the grid. However, the biggest winners will be those living in lower latitudes, who will have access to the most sun and will be able to sell this bounty to the large migrant cities further north.

So, power will be generated locally, but also transmitted around the world from where it's generated to northern cities in cables or transported in the form of a clean fuel, such as hydrogen.

Electricity grids produce, store and transmit energy, and they need to respond to daily and seasonal fluctuations in supply and demand – the sun doesn't shine at night, and the wind doesn't always blow. Hydropower has been more reliable and its potential in developed countries has been exploited over the past century. But large-scale hydropower dams have been divisive and environmentally destructive, and in many countries they are now being dismantled and river networks restored. In the US alone, some 1,600 dams have been removed, and thousands more have been listed for demolition in Europe. Hydropower is also becoming far less reliable in many places, leading to regular blackouts: some rivers are fed by disappearing glaciers, others are experiencing drought conditions, and extreme weather is destroying dam infrastructure, often with devastating consequences.

Solar panels floating on canals and reservoirs could generate more power to the existing hydropower turbines, while at the same time reducing evaporative losses. Big reservoirs, such as the High Aswan on the Nile in Egypt, already lose about a quarter of their annual water input due to evaporation. A 2021 study into the potential for hybrid hydro-solar power plants found that covering just 1 per cent of Africa's reservoirs could increase the continent's overall generating capacity by a quarter.[1]

Where hydropower remains viable over the coming decades, it is a brilliantly useful contributor to grids – the biggest globally by sector – providing 16 per cent of electricity. In the global south, a new dam-construction boom is under way, with at least 3,700 new dams proposed or under construction. They are all controversial because, despite promising electricity and development to some of the world's poorest regions, they risk huge environmental destruction,[2] as well as often producing significant methane greenhouse gas emissions. For instance, if the dams planned or under construction along the Mekong River are completed, some 96 per cent of the sediment supply to the Mekong Delta will be trapped, leading to erosion severe enough to threaten the very existence of the delta, which is home to 21.5 million people.

Access to energy is a key factor in whether people need to migrate, since energy can make unliveable conditions survivable for longer – by helping to provide cooling and water supplies, for instance. In the Horn of Africa, where climate emigration is already well under way,

Ethiopia is constructing the GERD dam, which will tap the Nile to produce 6,000 megawatts but displace 20,000 people. Downstream Egypt, which is battling its own climate crisis, faces severe disruption to crop irrigation for years while the dam is being filled. Energy is linked to food, climate and poverty – how these elements are resolved will determine whether populations stay or migrate.

The massive Grand Inga hydroelectric dam project planned by the Democratic Republic of Congo on the Congo River has a capacity of 40,000 megawatts and a projected cost of $90 billion. It could dramatically alleviate poverty – although it isn't clear where the electricity will go. The poor, local rural populations most impacted by large hydro dams, and who most need electricity, are seldom the beneficiaries. Many states sell the power to neighbouring countries, others will send it to their cities. People who want to benefit, then, would need to migrate to cities.

In the northern latitudes, where many of our expanding populations of migrants will live, hydropower is generally expected to become more reliable with climate change, and sustainable designs could help power towns there. Hydropower is the reason Norway has such low emissions, for instance. Small hydropower installations have little environmental impact and are excellent for remote communities. Tens of thousands of these are in use across Asia and some parts of Europe, and could be deployed far more widely across the globe relatively cheaply – nearly 400,000 potential sites for small hydropower have been identified in Europe alone. China has a total installed capacity of around 80 gigawatts in small hydropower, almost quadruple that of the Three Gorges Dam.

Another reliable source of energy comes from the heart of the planet, and this could be a game changer for new northern cities. Geothermal energy is already used in places where hot gases or liquids from the Earth's interior breach the surface through hydrothermal vents, geysers and fissures. However, the biggest potential is in tapping the heat of the Earth's rocks, deep down where they're not exposed. The scope is massive: this heat can be mined across the planet, theoretically at any location, and of course it's always 'on'. Perhaps the most promising technique in the short term is a deep 'closed-loop' radiator construction, where two wells are sunk 2.5 kilometres apart and connected laterally by a series of sealed pipes filled with fluid that rise up to the surface where the heat can be used. Because the loop is closed, cold fluid sinks on one side as hot fluid rises on the other, so there's no need for a pump. This easily scalable system is ideal for cities or other areas where land is in high demand. It can provide the baseload for a grid, and can easily be turned on and off as needed by restricting or cutting off the flow of fluid. A handful of experimental plants are piloting the technology, and in the next decade former oil industry workers will be employed to sink geothermal systems like these across many northern cities and towns.

The other reliable source of energy that doesn't produce carbon uses nuclear power – it's the reason that France has such low emissions, for instance. This could power energy-intensive industrial processes like steel-making, replacing fossil fuels. Large nuclear power plants, which release energy by splitting atoms, provide baseload for grids across the world – making up 25 per cent of all power in the EU (10 per cent globally) – although many are ageing and running into problems because of climate change affecting their cooling facilities. Replacing them and building new ones is expensive, particularly when compared to renewable deployment, and burdened with political and cultural issues that make them a hard sell. This may change.[3] Governments must invest in the infrastructure and expertise to drive down costs, and there needs to be an easy pathway for private-sector investment if they are to be financially viable. In the meantime, the first small-scale modular nuclear reactors are expected to come online towards the end of the 2020s and could prove a versatile source of clean energy. Russia is even designing a floating version that can be towed to wherever the power is needed, which could be very useful as the Arctic opens up to new industry.

The other sort of nuclear power is fusion, the energy that is released as two atoms are forced together to form a larger one. This long-hoped-for energy source has been something of a distant dream for almost a century; however, recent innovations in smaller, cheaper, modular technologies have made it a more promising proposition. The first fusion reactors could start entering grids by 2030, with the UK promising its first fusion power plant by 2040. To be clear, this is late – global emissions will have to have dropped by 45 per cent by

2030 in line with our net zero climate targets, but fusion offers the promise of energy that is plentiful and virtually free, which would transform our lives. Consider that every human activity on the planet requires energy, whether that is making food, clothes or toys, and today this energy is limited and polluting. If energy became plentiful, free and not polluting, it would transform our relationship with the planet. Arguably, it could make it worse (and there are many environmentalists who would like us to use less of everything no matter what). But it could make it much better. Today, some of the worst environmental destruction is caused by those who have the least, because energy poverty means people are forced to cut down their forests, pollute their rivers and coasts, hunt their wildlife and live and work in dangerous and dirty conditions.

In the meantime, large chemical batteries are being built across the world to store renewably generated power and release it into the grid on demand. It can also be stored as thermal energy in molten salts and as 'pumped hydro', in which surplus energy is used to pump water from a lower to higher reservoir, where it can be released to drive turbines as needed.[4] These storage systems will be essential to meet the huge energy demands of our megacities, particularly during winter when the sun sets early in the north and people will need power for lighting and heat.

Hydrogen, made by using renewable power to split water molecules, is another way of storing energy. Hydrogen can be compressed and shipped around the world, where it can be burned to drive turbines or used in a fuel cell (a sort of battery) to produce electricity. Australia is planning a large industry, effectively sending its abundant sun to cloudier northern climes by generating hydrogen to ship around the world in the more transportable form of liquid ammonia (another reaction is needed to recover the hydrogen). For Australia and other reliably sun-drenched nations, this energy-supply economy using a small labour force will be a good option after mass emigration.[5]

Most transportation of people and goods will need to be carried out on the ground rather than by air, because of the limitations of battery weight. Our northern cities will be linked by high-speed rail and shipping routes,[6] powered by electricity or, possibly, nuclear. Sail power is due a revival, particularly with smart AI-controlled sensors and adjusters that can ensure optimum wind capture, enabling sails to augment – and in some cases replace – other forms of power on ocean vessels.

As we move fully into the urban age, today's polluting cars cannot all be replaced with their electric equivalents – even if we could stand the gridlocked traffic of millions of private vehicles, the energy costs of the material resources involved are too great. Moving around our cities will be safer, healthier and quicker using electrified public transport and incarnations of the electric cycle rickshaw and cargo bike.[7] Where electric vehicles are necessary, they could be hired or shared. Charging, too, can be distributed – there are already apps that connect drivers to households with electric-vehicle chargers that they will share for a fee.

Flying is harder to decarbonize. However, it is contrails, the vapour trails made by planes, not carbon dioxide, that cause two-thirds of the climate impact of flying, accounting for roughly 2 per cent of global warming. Small changes to the altitude and time of day that flights are made can affect contrail formation, and they could be avoided entirely by rerouting planes to higher or lower altitude, at little cost and huge benefit.[8] Taxing aviation would help to reduce unnecessary business flights that could be replaced with video-conferencing, but the extremely wealthy in society, who cause, disproportionately, by far the most emissions, will not be dissuaded by reasonable taxes. Instead, private jets should be banned unless they are electric. In the future, synthetic aviation fuels made from captured carbon dioxide and green hydrogen could bring a renaissance in flight.

Airships, or blimps, could also have a role in our northern world, helping transport cargo to the cities and remote mines of the Arctic during the seasons when they are inaccessible by ground transport, and several companies are already exploring the idea.

The issue we face is that the world's clean electricity generation is not sufficient to meet our energy needs. We therefore rely on burning fossil fuels for the coming few decades. In order to achieve net zero emissions by 2050, rich countries need to stop burning fossil fuels by the mid 2030s, with coal and oil phased out globally before 2040 and gas soon after. Of the more than 400 climate scenarios assessed in the 1.5°C report by the IPCC, only around fifty avoid significantly over-shooting 1.5°C. Of those only around twenty make anywhere near realistic assumptions on mitigation options, for instance the rate and scale of carbon removal from the atmosphere or extent of tree plant-ing. And even these involve 'challenging' strategies that are either unproven at scale or socially problematic. Realistically, we are extremely unlikely to keep below 1.5°C.

While most people are now convinced of the need to urgently phase out coal, fossil fuel companies and governments – and many power companies are state-owned – are banking on us continuing to burn fossil gas (methane) and oil, and simply capturing and storing the car-bon dioxide produced, to prevent additional atmospheric warming. Sounds great, except carbon capture and storage (CCS) has never been used at scale, so we don't know if it would work effectively. Nevertheless, it's the plan. Whether governments and investors con-tinue to sponsor fossil fuel exploration and extraction remains to be seen, especially as vast new fields open up in the Arctic in coming decades, and the industry talks up the promise of CCS and other post-combustion mitigation measures. Even if CCS were proven, burning fossil fuels remains a filthy and damaging industry.

It is, by some understatement, expensive to decarbonize our econo-mies. Nevertheless, it must be done to transform our dirty, unjust global societies. Governments have thus far failed to approach the problem with the ambition and scale it requires – there has been nowhere near the level of commitment that is given to a war effort, for instance. One solution would be for banks to work together to create large amounts of money to fast-track the transition, in what's been dubbed 'carbon quantitative easing'. This, some suggest, could allow petrostates to be compensated for loss of fossil fuel earnings in a similar way that slave owners were compensated during the abol-ition of the slave trade, helping speed the end of the industry.

BETTER GROWTH

We've looked at some of the ways the world intends to meet growing demand for energy, but we can also reduce this growth in demand. The most obvious way is by making the energy we produce go much further. There have been huge advances in energy efficiency, particularly in industry and through domestic technologies, which have helped wealthy countries, but they need to be deployed globally.

There is another way to reduce demand, and that is to reduce growth: during economic recessions and other reductions in activity, such as the Covid pandemic, emissions dropped significantly. A number of environmental activists are calling for an end to growth or even for negative growth, pointing out that the usually defined 'healthy' growth rate of around 2–3 per cent of GDP per annum is environmentally unsustainable. Globally, the average rate of GDP growth is around 3.5 per cent per year, and the world's environmental problems are getting worse. However, it is not growth that is the problem, but environmentally and socially unsustainable growth.

Economic growth is an increase in the quantity and quality of goods and services produced per person over time. It is the engine that generated the plenty we enjoy today. But wealth remains unequally distributed around the planet, and while some places, such as Britain, have experienced growth for centuries, others, such as Chad, are still mired in deep poverty. Many of today's very poor countries were exploited and impoverished by colonial powers that did not allow their economies to grow – and rich countries have continued to harm the prospects of poorer countries in many ways, since their independence.[9] Active policy to address this, along with funding, must be used to urgently help governments reduce poverty through key investments in renewable energy, universal healthcare and education.

As we've seen, people live in poverty not because of who they are, but because of *where* they are; whether they happen to be born into a large, productive economy or not. Cities are more productive than rural subsistence farming; rich countries are more productive than poor ones. Citizenship bestows its own bonus (or penalty) on a person's income and opportunity beyond their individual luck and effort.

No matter how hard a farmer in Chad works, she is extremely unlikely to become as wealthy as a person born in Manhattan's Upper East Side. Just by relocating to the United States, she could triple her spending power (because incomes are higher even though prices are also steeper). That is why enabling and assisting migration is key to reducing poverty.

Naturally, not everyone in poor countries can or would want to migrate, and it shouldn't be the only route out of poverty as it is in too many societies. Poor countries must be helped by the rich world to grow and adapt their economies. Climate change will make many locations unliveable, as we've seen, for a proportion of the population, but the wealthier the country, the more it can adapt, and the longer people can stay. Economic growth – helped by remittances from émigrés – is needed to build stronger, resilient societies for those who do not migrate and for those who return over the century. Economic growth is how societies are able to provide people with the opportunities, goods and services they need for a dignified 'good' life.

GDP per capita is a broad measure that's useful for comparing countries with each other and over time. However, natural capital (the services and goods provided by nature) is not measured when calculating GDP, and environmental destruction often scores as a positive – a forest generates GDP when it is chopped down, for example. This is clearly unsustainable, and economists including Kate Raworth, author of Doughnut Economics, have proposed more appropriate metrics to measure economic growth in the twenty-first century, and tools such as carbon pricing. We value what we measure, and we need to find better ways of measuring the things that actually contribute to a nation's wealth, such as clean air, healthy soils, and dignified elderly care, which don't obviously contribute to GDP or incomes. Recall, also, that economic growth is the increase in the amount *and quality* of products and services over time. Moving from coal power to wind power, even if exactly the same amount of power is produced, is an increase in quality of power – air pollution is slashed, greenhouse gas emissions are avoided and wind turbines are safer and require less maintenance. This, then, is economic growth. If scientists find a way to cure cancer or eliminate malaria, that is economic growth. In other words, economic growth is not intrinsically predicated on an increase in unnecessary consumption or of pollution; we do not need to replicate all the growth patterns of the last couple of centuries, we can grow better with better policies. As technological innovation drives up productivity in sectors where it can, such as manufacturing and agriculture, the proportion of labour required in those sectors shrinks. Automation takes people's jobs and this has led to huge social difficulties – part of the answer is to tax the robots, as South Korea does. That helps put some of the productivity benefit back in the public purse.

It's clear that there is tremendous inequality within and between countries, with nearly 3,000 billionaires at the same time as the majority of the world's people live on very low incomes without access to even basic products. Global poverty has declined in recent decades, but Africa has seen the least progress, with 37 per cent of people in sub-Saharan Africa living in extreme poverty. To get to a stage where people are no longer poor and can access the goods and services they need, we need to go much further than reducing inequality. Countries need to see a rise in the average level of income – they need sustained economic growth.[10]

This puts us into problematic territory, though, because an increase in production demands an increase in energy and resources. In recent years, more than thirty countries have managed to decouple GDP from carbon emissions,[11] but most nations have not. Vast and rapid deployment of clean energy is the only way to meet this crucial growth, and it will also provide incomes. The IEA's road map projects an 8 per cent fall in the amount of energy used by the global economy by 2050, despite a doubling of GDP, a population rise of more than 2 billion people and the provision of universal energy access by 2030. This would be accompanied, the report finds, by increased employment in the green economy.

The only way for our massive new culturally mixed cities to work successfully is if they, too, manage sustained economic growth.

Smart, distributive policy is essential: emerging sectors from biotech to clean energy could provide huge opportunities for generations of native and migrant workers to build economically sustainable, fair societies – or they could end up concentrating even more power, wealth and opportunity in the hands of those few who already have unequal allocation. The post-industrial transition leaves the majority of jobs in sectors that cannot easily increase productivity purely through technological innovation, such as the service sector. We will always need people to work as hairdressers and nurses. By mid century, the UK will have a very old population, and the only way to meet its labour requirements will be through migration.

Universal healthcare and education are key to growth. Education is key to improving income, and it will be essential for most of this century's growth industries, from biotech to nanotech to material science. Education offers a route for migrants to cities with good economies, and that is why cities wanting to attract migrants need to have higher education institutes. Lack of access to healthcare is a huge economic burden, stealing lives and livelihoods. Most rich countries have universal healthcare, although notably the US has not. Other countries fail their citizens in other ways: consider that the UK has the worst statutory sick pay of any OECD nation (part of what exacerbated its Covid toll), and its elderly citizens are more likely to live in poverty. As migrants move to new cities, they will need access to healthcare and will be among the labour force providing it; it is essential that their countries of origin don't lose out in the process. Poor countries, many of which have substantial populations that lack even basic medical provision, need to be helped by rich countries to provide adequate facilities, particularly in fast-expanding cities, and to educate generations of healthcare workers.

The way wealth is distributed in societies must be transformed through policy, using tools such as inheritance tax, wealth tax and land tax. A carbon tax and water pricing would help preserve ecological assets, too. There is no justification for billionaires in societies where many struggle to feed themselves – this pathological accumulation of wealth could be far better used by society with no harm to lifestyle. We are still a long way from achieving economic growth without environmental destruction, but there are ways of getting there. Although some environmentalists argue we should be aiming for degrowth, I remain unconvinced that living standards could be maintained under such circumstances, and can't imagine democratic societies choosing a decline in living standards.

CIRCULATING RESOURCES

The advantage of having a global population concentrated in megacity safe havens is the opportunity for efficiencies. Community resources – including transport, children's toys, office equipment, heat, light and power – can be shared rather than owned, reducing raw materials used and waste. We already see the seeds of this sharing economy – I am a member of my local toy library and have ZipCar membership. Cities are not autonomous, they feed on the outside world. Swedish cities, for example, import 20 tons of fossil fuel, water and minerals per person per year.

With many of the sources of our minerals and natural materials set to be, within decades, physically inaccessible to humans – because of dangerous temperatures or rising sea levels – we will have to find alternatives or use robotic labour. (Implausible as this may sound, one major Japanese firm, Obayashi, is already constructing a massive dam using artificial workers in the southeast corner of Hokkaido.)[12] Meanwhile, new mineral deposits are already becoming accessible in northern latitudes, from the Arctic to the deep seabed, so analysts don't expect us to run out of supplies. But we must develop low-impact ways of extracting them that use less fresh water, no fossil fuels, and which do not create the pollution and ecological devastation that plague mines.

As societies become increasingly electrified, we'll need large quantities of copper and a periodic table of other elements, many of which were until very recently discarded as impurities when mining for other metals like tin. The IEA has warned that mineral supplies will need to increase thirty-fold by 2040 just to meet the demands of the electric car industry. This demand will lead to a resumption of mining in places where it had become unprofitable decades – even centuries – ago. Cornwall, for instance, is reopening its tin mines for lithium extraction, powered by geothermal energy from the same mines.

Greenland, with huge resources of thorium and uranium and rare earth metals, is being courted by the US, China and Australia, each racing for approval to begin mining there. The debate over granting approval triggered a political crisis in the world's largest island, forcing a general election in 2021, with Greenland's 56,000 citizens torn between protecting their fragile environment and developing their economy, which currently relies on fishing and grants from Copenhagen. On that occasion, the environment won.

Resource scarcity will force a move towards a circular economy, in which every product's end of life is considered at the design stage so materials can be easily reused and cycled continually with little waste. Effective low-energy plastic-recycling methods have been developed, that can turn any plastic back into oil, from which any plastic can be made. The scourge of plastic waste should finally end. New materials made from abundant resources, such as carbon, will also help improve the sustainability of product manufacture. Fast-growing materials, like bamboo, will be grown in accessible tropical locations, as well as in plantations further north.

The resource anxiety that will dominate this century, though, is water: chronically too little, and regularly too much. Ninety-eight per cent of the world's water is salty; most of the rest is held in glaciers in Antarctica and Greenland; eight-thousandths of 1 per cent is in all the rivers, lakes and wetlands combined; and just one ten-thousandth of 1 per cent is clouds, vapour and rain. So, most of the water we rely on is incredibly rare, and we don't store enough of it – despite building millions of dams, reservoirs and ponds, we store less than two years' supply. What's more, this water is not distributed evenly across the globe: while Canada, Alaska, Scandinavia and Russia have more rivers than they care to name, Saudi Arabia has none. Norway has 82,000 cubic metres of fresh water per person; Kenya has just 830. Some of the world's most important and mightiest rivers are so heavily extracted, including the Nile, the Colorado, the Yellow and the Indus, that they barely trickle to the sea.

Water issues will be the main driver of climate migration over the coming decades.

Today, around 4 billion people – two-thirds of humanity – experience water scarcity for a least one month a year. Half of them live in China or India. At least twenty-one cities in India, including New Delhi, Bengaluru, Chennai and Hyderabad, will run out of groundwater by 2025, affecting around 100 million people. And 40 per cent of India's population will have no access to drinking water by 2030, according to a report by the National Institution for Transforming India, the country's principal planning organization. Already, hundreds of thousands of people are regularly forced to queue for rations of water brought by tankers. Melting glaciers initially increase flows for mountain rivers . . . until they disappear. More than half of the world's important glacier-fed rivers have passed this 'peak water' threshold.

Globally, some 70 per cent of water use is for agriculture, but when water is scarce, cities always trump farms, which means devastation for farmers and increasing food insecurity. People will have to move.

Some rivers will receive greater precipitation, although not necessarily at the most useful time for agriculture. Rather, there will be an increase in flash floods, erosion and loss of crops and life in unpredictable events. European cities such as London, for instance, will experience periodic flash floods followed by droughts. This means that the annual precipitation arrives in unmanageable and destructive storms, which slide off streets and other hard surfaces into rivers, or evaporate without replenishing aquifers. Cities then have to contend with months of no precipitation at a time when it is hotter and people are in more need of water. Californian residents are already buying machines that condense drinking water out of the air, but these are energy demanding and costly.

City-scale underground reservoirs will need to be built even for our cities above the 45th parallel to harness and recirculate rainwater, as is already done from Singapore to Orange County, California. Unappealingly referred to as 'toilet to tap', when it is done effectively, with minimal leaks and evaporative losses, closed-circuit water recycling fully filters, cleans and stores the city's supply. Israel leads the world in water conservation, helped by a water tax: 85 per cent of purified sewage is recycled. Funding for this infrastructure can be generated by progressive water taxes, which will also shift people from wasteful behaviour.

New water policies will be increasingly important for residents as well as industry, even in places that historically haven't lacked supplies. These include curbs on hosepipe use, unsustainable golf courses, mandatory covers on water collectors to prevent mosquitoes, rainwater storage for roofs, water-saving home appliances, drain-clearing responsibilities and a ban on building in flood zones. Seawater desalination plants, powered by renewable or nuclear energy, will help coastal cities and irrigate local agriculture.

Canada's south-central prairies and the Russian steppes will become drier as the planet heats, something that river diversions could help to alleviate for agriculture. However, most of the rest of the northern region will become wetter, and migration will follow the water. By the 1990s, a greening of the northern Arctic owing to increasing plant cover was already visible by satellites. In Greenland, seal hunters have begun farming, and their potato harvests are so much better than elsewhere in Europe that Danish scientists have been studying them to learn why.

Water scarcity is likely to trigger conflict, which produces its own migratory displacements. Seeding clouds to make it rain artificially is likely to be expanded, for instance, despite concerns that clouds are being 'stolen' from neighbouring states. The UAE already seeds clouds regularly to fill its reservoirs, as does the US during times of drought; and even simply to improve skiing conditions. New reservoirs, river diversions and canals will also need to be built to direct water differently. These will also undoubtedly prove controversial, especially as a large number of the world's most important rivers are transnational, or flow from major 'water towers' held in a different nation. This means, for instance, that Ethiopia controls a water tower for Sudan and Egypt; the US controls a water tower for Mexico; and China controls a water tower for Bangladesh, Burma, Laos, Cambodia, Thailand and Vietnam.[13] China is hoarding vast amounts of Himalayan water behind huge dams, and has even created Chinese villages and deployed security forces inside the sovereign nation of Bhutan to secure its hydrological claims.[14]

Major new waterways and river diversions are planned this century to deal with water scarcity and boost development over the coming decades. If they work, they could reduce or delay the need for millions to migrate. The best known is China's immense South-to-North Water Diversion Project to deliver water from the plentiful Yangtze to the oversubscribed Yellow River, which is expected to be fully completed by 2050. Central and Western Africa's Transaqua Project intends to divert 50 billion cubic metres of water annually from the Congo River

Basin along a navigable canal – with a series of hydropower dams – to the Chari River, which flows to Lake Chad. The plan would provide hydropower and transport routes to the DRC, Republic of the Congo and the Central African Republic, as well as supplying water to the severely depleted Lake Chad, which has shrunk by 90 per cent in the past fifty years, with devastating consequences for the region. The diversion could irrigate up to 70,000 square kilometres of farmland in Cameroon, Chad, Niger and Nigeria, and feasibility studies for it are being funded though China's Belt and Road Initiative.

India's National River Linking Project is on a whole other scale. This is a plan to reconfigure the subcontinent's rivers to take water from dozens of Himalayan headwaters in the wet northeast to the arid lowlands, where existing rivers would be linked to each other through canals, reservoirs and dams. Critics point out that better water management would alleviate many of the country's water scarcity issues without the need for this extraordinary hydrological reconfiguring – the excavations alone would make it the single largest construction project ever carried out on earth – which carries huge potential for environmental (and cultural) damage. However, the $168 billion project, if completed, would transfer 174 billion cubic metres of water per year, generate 34,000 megawatts of electricity, and increase India's irrigated land area by a third.

Humans are not truly limited by scarcity of resources – we can always innovate past them or find new sources of whatever we seek. The limitations in power, water, minerals or wealth are, in truth, human limitations. The sun alone gives us enough energy in an hour to power our entire world for a year; water is all around us, it just needs desalinating. Everything we need and more can be made from the body of the earth and its biology. It is through our invented human socioeconomic system that we limit ourselves. We can innovate ourselves out of that limitation too, if we choose, but it is far harder to do. We may find we have no choice.

12

Restoration

Allow yourself a moment to grieve for the world we've lost, the biodiversity and culture we're losing, the time we've wasted not listening to climate scientists and green activists. The very idea that we need to make a plan for surviving an increasingly uninhabitable Earth through the horrendous upheaval of mass migration because we aren't maintaining the liveability of our wonderful, self-regulating biosphere, is the worst folly. It's a truly awful situation.

But we have work to do. Let's look at how we can make the world liveable again.

Our epic migration occurs in the context of an extraordinary crisis for humanity: global climate catastrophe, a biodiversity crash, and rapid human population expansion. The coming decades will be a vital island in time in our species' history, during which we will need to facilitate our great upheaval, and at the same time restore the world to a healthier ecological and climatic state. The better and faster we are at restoration, the fewer people will need to migrate and the more pleasant all of our lives will be on this planet.

The two biggest problems – both of our own making – are biodiversity loss and climate change. Fortunately, these are connected problems, and to some extent nature restoration and climate restoration can be done through similar routes. Biodiversity loss is mainly the result of the way we use land, but also of over-hunting and climate change. And, since climate change is a cause of biodiversity loss – for example, drought harms soils and forests – reducing global temperature would also bring biodiversity benefits. And it cuts both ways: restoring biodiversity helps suck up carbon dioxide and store it, which brings down global temperatures and thus reduces climate change. In other words, the immense job provides multiple wins.

NATURE RESTORATION

Let's start with biodiversity loss. Human beings and the human systems we depend on are entirely underpinned by the world's living systems. Today, one-fifth of countries are at risk of ecosystem collapse. By degrading soils, diminishing forests, losing coral reefs and poisoning rivers, we're threatening our own survival. Natural systems are amazing because biology is self-replicating, so these resources never run out, given the right conditions – the problem is, we're drastically changing those conditions. Forests, for instance, are noticeably suffering from the effects of fires and hotter temperatures – in many areas, trees are no longer growing back after blazes because the soil conditions have shifted to barely support shrubland. This is a global problem – forests are at risk from Africa to Asia, Europe to Canada. Across the Rocky Mountains, one-third of burned forestry is not regrowing. Since 2010, half of all trees have died in California's Sierra Nevada. In Alberta, half of existing forests could go this century. By 2050, tropical rainforests are projected to pump out more carbon dioxide than they suck up.

One positive effect of humans abandoning large tracts of the globe through urban and international migration is that it will allow some natural restoration of lost biodiversity. Places left alone by people return to a wild fecundity surprisingly quickly. Higher levels of carbon dioxide and the greater amount of rainfall that is projected for the tropics, while problematic for humans, is better tolerated by plants. Some forests, mangroves and grasslands will revive. Some animal species that depend on them may see their populations restored. Just as migration is a saviour for humans, so too for animals – what is pushing many species to extinction is not climate change itself, but the inability to migrate to safety because of habitat destruction or the obstructions of human infrastructure. We must provide protected corridors for wildlife to mobilize, and ensure healthy breeding populations.

These can take a surprising form: decommissioned offshore oilrigs, for instance, can be transformed into thriving reefs and important fish nurseries in the open ocean.

The total weight of all human-made infrastructure now exceeds the planet's living biomass. Just 2.8 per cent of Earth's land is intact wilderness.[1] Some environmentalists, inspired by the late biologist E. O. Wilson, are calling for half the planet's land to be conserved for nature. Given the increasing global population, this is an ambitious target. However, if one-third of the planet's most degraded areas were restored and those ecosystems still in good condition were protected, this would prevent 70 per cent of extinctions and store the carbon equivalent of half of our emissions since the industrial revolution. Ecologists have mapped Key Biodiversity Areas, places that are important carbon stores, and other sites that are useful to protect (for instance as wildlife corridors). Together this 'global safety net' proposes going beyond the 15 per cent of land already under some sort of protection and adding a patchwork of other protections.[2] It is perfectly possible: Costa Rica has gone from being the most rapidly deforested place in the world to protecting one-third of its territory – and making money from its enriched ecosystems through nature tourism.

It is important to remember that indigenous populations live in many of the areas that have been identified for conservation, with their own needs and livelihoods. We can only protect wildlife if we also protect people. In some cases, this can be by paying communities to protect forests and wildlife. Recent inhabitants could also be compensated for leaving important ecosystems, through schemes that provide livelihoods and homes elsewhere. Financial tools to encourage markets to invest in restoration (rather than destruction),[3] will also help, perhaps including carbon quantitative easing.

Extinction rates are at least 1,000 times higher than they would be without human activities, and in some cases local losses directly threaten us. Pollinators, for example, are essential for many of our foods, yet in the UK 97 per cent of meadowlands have been lost through intensive farming, causing a crash in insect and bird populations. As we work to produce more food in restricted plots for a global population of 9 billion people, the impact on wildlife has the potential to become even worse. Policies to address it could include mandatory wildflower strips in fields, which encourage pollinators and cut pesticide use with little loss of cropland. Cities can play a role too. Private gardens in Britain cover an area larger than all of the country's nature reserves combined – over 10 million acres; these could be seeded with important flowers. Grass verges lining roads and in cities could be diversified and left unmown – again, in the UK, the cumulative verge area is equivalent in size to the county of Dorset.

Given the scale and extent of biodiversity loss, interventions are now essential in many cases to enable species to cope with human changes to the environment. We have genetic tools to help species adapt to Anthropocene conditions, although it's a time-consuming and expensive process. American chestnut trees in the US have been genetically altered to protect the species from blight; black-footed ferrets are being cloned to bring the species back from the brink; and coral colonies are being genetically engineered to withstand hotter seas. In some cases, species in danger will need to be helped to migrate through active translocation; in others, special protective zones and policies will be needed – in Rwanda, for instance, tourists pay a premium to visit gorillas and this money pays for community development projects as well as wildlife protection. As populations move away from the tropics, there will be plenty of roles for globally important nature guardianship and restoration for those who remain.

The next few decades of urban migration will see massive abandonment of rural areas, with farmlands across the global south consolidated into larger holdings and farmed much more efficiently, allowing unproductive marginal land to return to wildland. Genetically modified crop varieties should also mean fewer ecologically harmful chemicals are required. Vegetables that today are farmed using excessive pesticides could be grown in efficient, automated urban vertical farms without the need for pesticides, or by small family farms that focus on sustainable methods and are paid for maintaining and restoring biodiversity. Regenerative agriculture will help restore and maintain soil carbon and fertility.

Nature-based solutions to our environmental problems can only go so far – especially if the environmental problem is too extreme for the nature to thrive. 'Green wall' tree-planting projects across Africa and China have been building for more than a decade, aimed at holding back increasing desertification owing to global heating, but have shown mixed success. Many of the planted trees don't survive.

Tree-planting is very popular as a way of offsetting carbon emissions, but in many cases, the species chosen are not appropriate for the local conditions and could actually increase carbon dioxide emissions;[4] pasture may be better. In other cases, what is being created is a plantation desert with little biodiversity and few of the ecosystem benefits of a natural mixed forest.[5] Using trees as an 'emissions offset' brings additional issues, including double-counting of forests by multiple parties, lack of verification and long-term care of the carbon-suckers, and so on. If trees are planted to offset carbon emissions, it may take decades of growth to achieve this, and what happens if that forest burns down? It is very clear that the market and governance of offsets needs to be well regulated, and separated from emissions taxes and pricing.

Nevertheless, restoring the planet means replanting vegetation. In places that have been deforested through bad land management and remain viable for forestry, such as lowland Britain, tree planting could be hugely successful, and with investment should provide jobs. At the same time, we need to protect other key vegetation, such as grasslands in steppe and semi-arid regions, where it acts as an excellent carbon store and is far less likely to be lost to fire than inappropriately located forest. Peatlands too are vital, holding twice as much carbon as forest – peat is 50 per cent carbon. These important bogs are being logged, drained and burned for agriculture at an alarming rate from the UK to the tropics.

Sea grasses, mangroves and marshes are excellent carbon stores – thirty times better than terrestrial forests – and also help to reduce erosion and nurture fish and other sea life. Most of this type of ecology – sometimes referred to as 'blue carbon' – is threatened, and better protection and restoration of it would have multiple benefits as well as providing jobs. There are various projects around the globe attempting to cultivate and plant sea grass meadows at sites too deep to be disturbed by boats, and to reseed tropical mangroves, particularly in places where they have been destroyed for development. Kelp is another important, fast-growing carbon store with multiple ecosystem benefits – and, of course, people can also eat it.[6]

Restoring the planet's biodiversity is a global, labour-intensive undertaking that will provide useful jobs for migrants and natives, paid for by public–private partnerships. Such important 'global community' work could form part of a national service community-building programme in many new cities.

Not everything can be restored. Coral reefs, which support a quarter of all life on earth, and the livelihoods of one billion people globally, are not expected to last more than forty years, because they are killed by heat and acidic conditions. At 2°C of warming, coral reefs 'mostly disappear'. However, even achieving 1.5°C will result in the loss of 90 per cent of reef-building corals compared to today. As a diver, I find their loss heartbreaking, but their importance goes far beyond beauty. Currently, reefs contribute an estimated $10 trillion in ecosystem services, including protecting vulnerable coastlines from erosion and storms, and producing the sand of our beaches. Where we cannot protect ecosystems in their current state, we will need to try to restore some of their functionality, for instance by creating artificial reefs to support fish nurseries. Genetic engineering and selective breeding of coral polyps and their algal symbionts to produce variants that are more heat-resistant may help prolong the viability of a few reefs – and eliminating other stresses, such as pollution and boat damage will also help – but unless the global temperature stops rising, reefs are doomed.

CLIMATE RESTORATION

About half of the world's population will live in vulnerable tropical regions by 2050 (up from 40 per cent today), but much of this zone will start becoming uninhabitable before that date. If we turned down the temperature of the planet, fewer people would be forced to migrate, and those who have been displaced could return. However, the methods for doing so, known as geoengineering, are mostly untried and controversial.

One way to do this is to reduce the amount of carbon dioxide in the atmosphere, which is difficult to do at scale (considering we're still adding carbon dioxide) and slow, but well worth trying because it would restore the safer, more stable climatic system, and in many cases improve biodiversity too. The other option is to physically cool the planet by reducing the amount of solar radiation that's heating it. This could be done through techniques such as injecting reflective particles into the stratosphere.

Our hands are on the planetary thermostat, so we have the choice of whether to start limiting warming early on, before millions of people need to move, or whether to wait for some later crisis, such as back-to-back heatwaves that kill thousands, before deploying geo-engineering techniques. This effort to limit global temperatures requires an unprecedented political, social and technological response. The stakes could not be higher. For if the world becomes much hotter, even migration will not save us.

Obviously, we need to stop adding to the problem, which means not burning fossil fuels and preventing the world's soils from releasing their carbon stores when they're broken down during agriculture, drought, deforestation or heat. We also need to remove from the atmosphere the carbon we've already emitted, through restorative planting and making sure the carbon that is removed is locked away. Forests do this so-called 'negative emissions activity' until they are chopped down or burned; sea grasses are excellent at locking carbon down into the seabed. Another strategy is to take vegetation – and this could be corn stubs or other agricultural waste – and heat it in the absence of oxygen to produce charcoal ('biochar'), which is a solid form of carbon that can be buried in soils to enrich them. The land can then be used to grow more vegetation, withdrawing more carbon each time. The trade-off is that this agricultural waste cannot be fed to animals, mulched or used for other purposes.

Bioenergy with carbon capture and storage – usually called 'BECCS' – is extremely popular with governments and corporations hoping for cheap negative-emissions technologies to offset their emissions. BECCS involves growing plants for fuel, burning them in power stations and capturing the carbon dioxide produced, which can then be stored safely. The problem is that the amount of land needed to make a significant difference to our net global emissions is vast – by some estimates as much as 80 per cent of current cropland – and as we've seen, it is needed for food and wildlife. To use precious land to grow a fuel just to burn it is obscene.

Ocean fertilization offers excellent potential for climate restoration without taking precious land. The oceans are naturally fertilized by minerals blown off desert soils, and these nutrients enable the growth of phytoplankton, which absorb an estimated 40 per cent of all carbon dioxide (four times the amount captured by the Amazon rainforest) during their photosynthesis. Phytoplankton are the basis of the oceanic food chain and thus hugely important for biodiversity. Their growth is limited by lack of nutrients, especially iron. Adding powdered iron to ocean waters, in locations such as Antarctica, would dramatically increase phytoplankton production, sucking up carbon dioxide, and thus also reduce ocean acidification. In past geological eras, far greater quantities of desert dust were blown into the seas, resulting in global cooling. Artificial fertilization of the oceans would have the same effect.[7]

Large oceanic mammals, especially whales, help with this service. Whales feed in the deep ocean, returning to the surface to breathe and excrete iron-rich faeces, creating the perfect growing conditions for phytoplankton.[8] Industrial whaling in the twentieth century devastated this carbon-rich marine ecosystem, and protecting whales would significantly help rectify this.[9] Assisted ocean fertilization would help restore whale populations by supporting larger populations of krill, which feed on phytoplankton and which, in turn, are the food whales depend upon.

Restoring this complex cycle through ocean fertilization is a better prospect for reducing global temperature by sucking up carbon dioxide than much of the planned, unfeasible terrestrial tree-planting, yet it is currently classed as 'geoengineering' and thus regarded as too risky an intervention and not sanctioned apart from small-scale scientific experiments.[10] One of the fears with ocean fertilization is the uncontrolled growth of algae, which depletes shallow waters of oxygen, killing off other marine life and creating 'dead zones'. This is what happens when agricultural fertilizers are overused and end up polluting terrestrial water bodies and coastal zones. However, fertilizing the ocean in places where nutrients are limited and the circulation is strong removes the risk of dead zones forming, and could have the added benefit of helping to clean and de-acidify the water for essential shell-building organisms, such as plankton and corals. Pilot studies are under way, but larger restoration should be tried now.

Carbon capture and storage (CCS) technology will need to be installed in every power station that burns carbon. With most power station chimneys today releasing a fairly concentrated mixture of about 10 per cent carbon dioxide, there is huge scope to stop the continued burning of fossil fuels from adding to our global heating crisis. There is ample storage capacity for this carbon, in deep saline or sediment-filled aquifers. Some can also be sold for use by industry, in commercial greenhouses, or combined with hydrogen to make synthetic fuels. Today, where this technology is deployed, the carbon dioxide is mainly pumped into depleted oil wells to enhance the extraction of more fossil oil or gas, which is far from ideal. The cost has held back pure storage in most countries, but with carbon pricing and binding net zero targets, its deployment is unavoidable – and scaling up CCS will, of course, reduce its costs. Government investment will be essential.

We can also capture carbon directly from the air geologically by enhancing the natural process of 'weathering'. This is the continual erosion of rocks as they chemically react with the carbon dioxide dissolved in rainwater, and then crumble off, largely washing into the ocean, where the carbon is stored in the seafloor. This process absorbs around 0.3 per cent of our carbon emissions, but by enhancing it, we could dramatically improve on this. Some rocks are better than others. Silicate rocks, such as basalt or olivine, which are common on the Earth's surface, can be crushed into a highly reactive powder that can be spread on agricultural fields, where plant roots and microbes in the soil speed up the carbon dioxide removal. The powder is also an excellent way to add minerals to soils, boosting nutrient levels,[11] improving crop yields and helping restore degraded agricultural land. Silicate rocks also improve the health of crops, and protect them from pests and diseases. Since enhanced weathering makes water more alkaline, it can help counteract the acidification of soils through the overuse of fertilizers – farmers often add limestone to soils for this purpose, but silicates could be used instead. Together, these benefits would increase the profitability of farms and encourage the agricultural sector to take up enhanced weathering. All this, while at the same time removing carbon dioxide from the atmosphere.

Enhanced weathering can also be used in the oceans – silicates can be spread on beaches and washed into the sea on tides. This would help remove carbon dioxide from the oceans – meaning the oceans could absorb more of the gas from the atmosphere, reducing global temperatures – and it would help reduce ocean acidification, especially close to the areas the silicates are spread. It could prove a lifeline for coral reef ecosystems.

The problem is cost: mining and pulverizing rock and spreading it around on a large scale is energy demanding and expensive[12] – costs are several times greater per tonne of carbon dioxide than for BECCS, for instance.[13] But there is huge opportunity for the mining and oil and gas industries to invest in the process as an 'offset' for emissions elsewhere; and mine tailings or residues, which today pose a disposal problem for the industry, could be used for enhanced weathering in a win-win situation for everyone.

A more popular option is to use CCS-type technology to capture carbon dioxide directly from the air, and there's huge investment in this with multiple start-ups creating big boxes to do so. The problem is that in the air, the gas is only in concentrations of 0.04 per cent, so removing it takes lots of energy – potentially half of today's global energy supplies, according to one study.[14] Also, there is a *lot* of air, so this would require a global industry, sucking air in at vast scale, and capturing the carbon dioxide for storage. Nothing even approaching the scale envisaged exists. The process of removing the carbon dioxide would require millions of tonnes of solvent and considerable energy,[15] making direct air capture (DAC) extraordinarily expensive, and resource- and energy-intensive. Even if it were successfully deployed at scale, and carbon dioxide concentrations significantly reduced, there is a further issue: the oceans might respond by returning some of their carbon dioxide back into the atmosphere. Carbon dioxide is in constant flux between the atmosphere and oceans, and no one knows how this new disturbance to the equilibrium would play out, but scientists calculate that as much as one-fifth of carbon dioxide removed by DAC could be replaced by ocean release.[16] Nevertheless, it is essential we try direct air capture at scale, as part of our increasingly hopeless quest to stay below 2°C temperature rise.

Concerningly, all of the official road maps to net zero by 2050 rely heavily on one or both of BECCS or DAC – neither of which has been proven at scale. While on some levels betting on future technology may be more realistic than betting on consumption reduction through massive societal change, I am far from confident either will prevent us from exceeding dangerous global temperatures.

We are already at a global temperature rise of 1.2°C and suffering: in the past decade, 21.5 million people per year have been forced to move because of extreme weather, which is three times as many as were affected by conflict and nine times as many as were forced to flee persecution – and this is expected to triple in the coming decades. In 2020, the world incurred $210 billion of weather-related damage. When global temperature rise exceeds 1.5°C (which could occur by 2026),[17] some 3 billion people will be living in places that regularly experience conditions beyond the human habitable range. My feeling is that the world will wait some decades before reducing global temperatures, by which time mass migration will be inevitable.

Given what's at stake, I've no doubt we will start deploying tools to reflect the sun's heat away from Earth to maintain global temperatures at safe levels. Today, this form of geoengineering – deliberate and large-scale intervention in the Earth's climate system – remains taboo in a way that increasing temperatures via emissions is not. Let's be clear: planet-scale land-use change, vast atmospheric pollution, the release of massive stores of fossil carbon that heats the atmosphere and oceans ... these certainly amount to geoengineering even if it is not labelled as such. In our attempts to restore the planet to a more liveable state, we should be using all the tools we have.

Ice loss is accelerating at a record rate, with Greenland and Antarctica melting fastest. This is already raising sea levels, and a lot more of this ice melt is inevitable because of the emissions we've already released. There are various proposals for dealing with catastrophic ice melt by increasing its albedo (reflectivity), such as coating glaciers with reflective fleece blankets – this has thus far only been used in the European Alps. Another idea, currently being trialled in Alaska,[18] involves blowing artificial reflective snow made of silica (glass) onto glaciers. This increases the reflectivity of the ice by 15–20 per cent. Massive deployment of this, costing an estimated $5 billion, could reduce temperatures there by 1.5°C, and increase ice thickness by up to 50 centimetres, buying fifteen more years of warming time, according to the project's organizers.

Researchers have also suggested using wind-powered giant pumps to refreeze the Arctic sea ice. During winter, the pumps would draw seawater out of the ocean and onto the surface of the ice, thickening the existing sea ice, so that it won't grow so thin during the warmer months. Glaciologist Michael Wolovick of Princeton University has suggested that some Greenland or Antarctic glaciers could be potentially stabilized with the use of an artificial sill – a kind of mound on the ocean floor, perhaps constructed out of sand or rocks – that would block warm ocean water from seeping beneath the ice and melting it from the bottom up. Another idea is to brighten low-altitude clouds above these important polar glaciers by spraying them with droplets of salt water, using a fleet of remotely operated drone ships, thus helping them to shade the ice and reduce its temperature. The former UK Chief Scientific Adviser Sir David King is spearheading a project that hopes to try this in 2024.

The same cloud-brightening idea could also protect coral reefs from bleaching in hot temperatures. Scientists have carried out a trial above the Great Barrier Reef, using a modified turbine with 100 high-pressure nozzles to spray hundreds of trillions of nano-sized ocean-salt crystals into the air per second, from the back of a barge. The equipment boosts the natural process of cloud formation, which mostly occurs when moisture gathers around salt crystals stirred up by winds from the ocean surface. The salt crystals in the experiment remain in the air for only one or two days, but there are plans to scale it up at least ten times, using more and larger turbines. At that level, it could generate clouds to cover an area of hundreds of square kilometres – enough to slightly cool ocean temperatures. Other options being explored to prolong the existence of living reefs, include spraying jets of cool water into the surface waters, creating a fine fog or mist above them; and, more feasibly, deploying a very fine surface film of reflective calcium carbonate. Similar technology is currently used to prevent evaporation in reservoirs and dams, including those used for drinking water, and it can reduce light penetration by more than 20 per cent. The surface films need only be applied periodically (by plane, sea vessel or automated buoy) during summer when coral bleaching conditions are predicted, to reduce solar radiation reaching the corals. It is likely that, in coming decades, this thin-film technology will be used more widely and beyond reef protection – increasingly to help conserve water in reservoirs, along with floating solar.

The worst impacts of global heating are felt where the people are, of course: on land. That needs a regional or global cooling technique, but it could be done by spraying a fine mist of sulphates into the stratosphere. While carbon dioxide heats the atmosphere by absorbing heat, sulphates cool it, because they reflect some of the sunlight back to space. The cooling effects seem to be more pronounced in the tropics than at the poles, so this wouldn't do much to protect us from sea-level rise owing to polar ice melt, but it would save potentially thousands of lives during heatwaves in the coming decades, and could prevent millions from needing to migrate.

The taboo over this kind of geoengineering has prevented even the most trivial tests being done in a transparent way, so we rely on modelling studies. Nevertheless, the chemistry and physics of the idea of deploying sulphates into the stratosphere is well understood in principle, and there is little reason to suspect that it wouldn't work to cool the planet. However, we still don't know how best to deploy sulphates, how regularly they would have to be released, nor what other atmospheric or climate impacts there might be, particularly in terms of key weather circulation and rainfall.

Cooling the planet would prolong the viability of heavily populated parts of Asia and Africa, enabling cities to use fewer resources on air-conditioning and water supplies than otherwise. Cooler, less evaporative soils would also make reforesting the tropics easier and more successful, helping to remove carbon dioxide from the atmosphere. Cooling would enable farmers to work outside without expiring from heat stress. It would also help their crops to thrive. The most effective way to protect crops against the worst effects of global climate change is to reduce the surface temperature – studies show this is more important than rainfall. In fact, cooling the planet with solar geoengineering would be more beneficial for crop yields than reducing carbon emissions, a 2021 study found,[19] because carbon dioxide is useful for plants during photosynthesis.

This is not to suggest, of course, that continuing to slash our carbon dioxide emissions, and the urgent removal of carbon dioxide already there, should in any way slow. The reverse: we only need to consider these expensive cooling techniques because we're continuing to heat the planet with carbon dioxide. Solar reflection wouldn't address some underlying problems resulting from too much carbon in the atmosphere, such as ocean acidification, but it would buy us more time to decarbonize and achieve negative emissions. Crucially, keeping the Earth cooler for longer would help the poorest people to adapt and alleviate poverty, which is vital morally and to restore ecosystems. Meanwhile, the effective capture from the air of today's carbon emissions can be scaled up. Everything we need to do to move into a net zero global economy, the increase in biodiversity we are aiming for, the improvements to people's lives and well-being ... everything would be easier in a world not also burdened with a catastrophic climate, frequent extreme weather events, drought and heatwaves. To have the option to cool the planet but not use it would be morally indefensible.

Sulphates – there are other options, but sulphates are the most promising – could be released into the stratosphere in a steady stream by high-flying aircraft or autonomous drones. The cooling effect would be immediate, but they don't last long in the atmosphere, so would need to be continuously applied and phased out in concert with falling concentrations of carbon dioxide – otherwise, the full impact of carbon dioxide heating would resume.

We haven't tried to cool the planet in this way yet, so we don't know whether there are any unwanted effects, and if so, whether they outweigh the effects of continued heating. However, the effects could be stopped with relatively little delay, simply by stopping the emissions. We don't know what effects a slight reduction in ultraviolet light may have on crop production and natural ecosystems,[20] for example. And there is a fear that some places might experience a reduction in precipitation with the technique. To address this, scientists have looked at the effect of modulating the process: by halving (rather than completely cancelling out) global heating with sulphate cooling, the models found the sulphate injections still offset most of the carbon dioxide-induced increase in simulated tropical cyclone intensity, and neither water availability, extreme temperature nor extreme precipitation are exacerbated in any region.[21] Generally, sulphate cooling would reduce drought – one study finds a reduction in the frequency of heatwaves and consecutive dry days in almost every region, relative to a climate in which atmospheric carbon dioxide is double the preindustrial level (expected around 2060).[22]

Nevertheless, adding increasing amounts of carbon dioxide to the atmosphere has certainly produced unwanted consequences with vastly unequal global impacts; we must ensure that we do not repeat this mistake. For instance, people adversely affected by solar cooling must be compensated. There needs to be governance and oversight. There is a Solar Radiation Management Governance Initiative, but it lacks authority and has no clear mandate.[23]

There are those who believe that if we take the 'easy' option of cooling by geoengineering, we won't put in the same effort to reduce carbon dioxide emissions. There is certainly no shortage of emitters trying to avoid or delay decarbonization commitments using carbon offsets as a tactic, so regulators must take precautions that geoengineering is used in addition to, not instead of, emission reduction.

There is also a moral queasiness from some sectors about using technology to fix the global heating caused by our 'bad' polluting behaviour; a puritanical zeal in insisting people drastically curb their lifestyles, for instance advocating negative growth, as a – perhaps morally better? – way of cooling the planet. For sure, there are individuals with obscenely polluting lifestyles and we all need to be mindful of our environmental impact, but some environmentalists seem much more willing to impose very large societal changes, with all the discomfort that entails, than make changes to the Earth's 'natural' state through geoengineering. This to me feels morally questionable, but that's the thing with morals, it's all subjective. For me, the morally right thing is to do whatever we can so that our fellow humans can live in a safe climate where they have enough to eat. This will mean helping those living in danger and hardship to migrate to safety; and reducing global temperatures so that climate stability is restored. (Remember that the temperatures we experience today are due to the bad stuff done in the past. The bad stuff we continue to do will show up in future heating.)

The multiple benefits of a green economy in terms of health, livelihoods, ecosystem protection and cost make ending fossil fuel combustion both urgent and the obvious way to reduce global temperatures, and the continuation of this process must be ensured by laws. But fossil fuels are embedded in our human systems, in food and energy production, transport and industrial processes, so replacing them is a slow and costly effort, made slower by nefarious interests. The process is slower than the physics of global warming, and billions of people are threatened by a likely temperature rise of 3 °C to 4 °C. This to me is an unacceptable risk. It means all efforts for cooling must be considered, with the more feasible all propelled forward. Sulphate cooling is certainly feasible.

Even with our best efforts to cut greenhouse gas emissions, we will still be heading for at least a 3 °C temperature rise. By injecting about 10 megatons per year of sulphates into the stratosphere, we could reflect away about 1 per cent of sunlight, keeping temperatures below 1.5 °C of warming. That might be enough to avoid catastrophic sea-level rise, limit drought, forest fires and hurricanes, and give some coral reefs a chance. For comparison, today's industrial pollution includes about 100 megatons of sulphates a year.

There are aspects to solar cooling which are less discussed. For one thing, the effects will be unequally distributed. Whereas global heating disproportionately hurts tropical nations and even produces some benefits for cooler countries, the opposite is true for cooling – the cooling effect of today's sulphate emissions from industry and shipping actually benefits tropical economies while harming cooler economies.[24] The same is projected to be the case for stratospheric global cooling. In other words, the tropical belt in the global south, which is home to billions of people, would benefit from the intervention, with improved crop yields and more liveable conditions, whereas northern areas, a few of which are already benefiting from warmer conditions, an increase in ice-free lands and better crop yields, would not significantly cool. What this means is that as carbon dioxide mitigation and withdrawal improves, and we reach net zero globally – and then begin the reduction in carbon dioxide concentrations such that we achieve 425 parts per million, or even return to 400 parts per million, northern regions will lose the benefits of hotter temperatures, while the global south will become less hazardous and more pleasant.

Geoengineering offers us the ability to choose the temperature of our planet – and we may not agree on what that ideal temperature is. People living in the tropics may well prefer a cooler temperature in which air-conditioning is unnecessary, and drought is rarer; whereas people living in northern latitudes may prefer a hotter temperature, particularly once we've adapted infrastructure and built new, thriving cities there. Living in London at 1.2°C of global temperature rise, I am now able to keep citrus trees in my garden, and I spend less on home heating. I enjoy the newly Mediterranean climate of southeast England much more than being cold. At the same time, this global heating has caused a five-fold increase in the number of extreme weather disasters over the past fifty years, which have killed over 2 million people and cost $3.64 trillion in economic losses.[25]

Over the past 4.5 billion years, Earth has swung between extreme temperatures, but for most of the time it has been hotter than today. In fact, mostly, it has been ice-free. However, our species evolved into the Pleistocene world of ice, and our civilizations were only created during the very recent period of climatically stable and relatively warm temperatures of the Holocene. Now we are pushing the planet into warmer temperatures, and pushing our own adaptability to its limits. But what temperature would we choose – pre-industrial average? 0.5°C hotter? 1°C hotter? – and who would get to choose? These are key questions for a global governance body, and it should be urgently appointed, with powers.

Restoration of the planet's biodiversity and climate to make it liveable for us and for wildlife would end much of the upheaval. The sooner we do it, the less mass migration there will be, with fewer people having their lives turned over. To be clear, migration won't end – it's a part of what it means to be human – but it will be far easier to manage and will, I hope, be managed well. If we don't get a grip on planetary restoration soon and at scale, then temperatures will rise to levels where nowhere will be safe for our human population.

It's not too late. Global population is expected to peak at 9.7 billion around 2065. From that point, it will begin to plateau or, more likely, reduce, returning to current population levels as early as 2100. This population contraction will remove the extreme pressure on our resources, although it will come with its own demographic concerns.

If environmental conditions allow, humanity's great migration will not end with the move to higher latitudes and colonization of the polar regions. Rather, it will continue into the next century, as people repopulate abandoned regions. In time, as we continue our restoration, humans will once again expand from their refuges to the far reaches of the planet.

Conclusion

It is absurd that we are considering the mass migration of billions of people. It's absurd that we are continuing to heat the planet, knowing the consequences. After a career spent writing, researching and talking about climate change and its impacts, I can't believe we're still in this situation. And yet we are. My daughter asks me, with the cool rationality of a six-year-old, why we don't just stop burning fossil fuels? 'It's not that simple, darling,' I reply, crossly.

It is that simple, though. We know the science, we have the technology – it would be very costly in the short-term, but we have the money and spend it freely on other things. We have made it complicated. We are trapped in the socio-political-economic web that we've woven. This invented trap, this human construct, is keeping us in so much danger that we are now in the absurd position of having to save humanity by moving vast populations of our species from places where humans have lived for hundreds of thousands of years.

Migration is inevitable, often necessary, and should be facilitated. But a situation in which billions of people are forced to leave their homes because parts of the world have been made unliveable is a tragedy. To a degree, this situation is not yet inevitable. We can reduce global temperature rise, and we must, which would avoid more extreme levels of human displacement. However, at 1.2°C, there is already considerable reason for people to migrate. We must stop seeing the people who move as the problem, even if some of their reasons are.

Much of this migration is to be celebrated: people moving to different cities, countries and continents enrich themselves and the societies they live in. Migration has been a fundamental part of our species' remarkable success story and the diversity and complexity of our cultures. If anything, it has been underused in recent decades, and many more people could gain from leaving their home towns and seeking their fortune elsewhere.

Migrants are the bridges between cultures, helping us to better understand ourselves as well as each other. My own experience of living in different places has given me important friendships, insight into other ways of thinking and living, intimate understanding of neighbourhoods and people, and new language skills. Just as important to me though have been the migrants I've got to know. Migrants are often individuals of courage, curiosity and determination who have taken a leap into the unknown, leaving familiar people, place and language, sometimes overcoming considerable hardship. Moving abroad isn't morally bad; we don't need to punish people for relocating. We make migration unnecessarily hard and often dangerous especially for those for whom we should make it easiest.

The social inequities of this world are baked in: we inherit our citizenship, we inherit the safety of our family home, we inherit our life chances. There's not a lot we can do about the triumph or tragedy of our birth circumstances. But safety shouldn't be left to the whims of inheritance. It is too important. We must make it easier for people to move around our shared planet – the planet that we have all inherited, unwittingly.

The mass migrations this century will be dominated by people from the poor, climate-ravaged world moving to the richer world, countries whose wealth has largely been enabled by changing the climate. This then is an opportunity to deliver some social justice, while benefiting both host and migrant populations with new growth. Cities need migrants in order to flourish, but immigration must be properly managed and supported – it needs to be a cooperative effort. There is great efficiency in cooperation, which is why evolution tends towards favouring it. This means global agreement on safe, legal pathways for migration, and mechanisms to share the upfront social economic costs of a large influx of new citizens. It's astonishing there isn't a coherent strategy of matching people globally to jobs, education and housing – something easy enough to do in our digital world. It is time to create one.

Migration is framed as a security risk for the developed world, which is wrong and needs to change. As I write, some 20,000 children are being held in detention camps in appalling conditions in the United States. These asylum seekers are freezing, hungry, and riven with lice and scabies. Europe, meanwhile, is testing a 'sound cannon' to be used against refugees, as part of a €37 million border control programme involving AI interrogation units and drones. Even while the EU heroically welcomed millions of refugees from war-devastated Ukraine, its response to asylum seekers from other places has deteriorated to the extent that it has begun criminalizing search-and-rescue efforts for migrants crossing the sea from North Africa.[1] The time, money and bad will that is being thrown (fruitlessly) at trying to stop migration could be far better spent managing and preparing cities to add new citizens to the labour force, and helping them to enrich us all. This means changing the narrative around immigration and forging an inclusive, strong and vibrant national identity. People who at once exaggerate and do down their own culture – ascribing it enormous value while having no faith in its ability to win over foreigners – should be shown this contradiction. Culture evolves and is enriched through complexity, not through static preservation.

There must be a pathway to dignified and safe migration for all who need it. It is inexcusable that migrants die every day attempting unsafe border crossings because they lack alternatives.[2] Facilitating migration is a key climate adaptation and will also address poverty and hunger. Investment in people in the global south, as well as in the large numbers of people who will move to and within the global north, is an investment in our shared future. It can start with a sharing of citizenship.

It's easier for people whose lives are relatively comfortable to appreciate the benefits of migration; poorer people in 'left-behind' towns can be fearful of the impact of immigrants. Given the mass immigration that is coming as people are displaced everywhere, it is the job of political leaders to reassure them – not stoke these fears – with policies that reduce poverty and create housing, services and jobs for everyone.

This coming mass migration has been clearly signposted for at least a decade, as we have accelerated towards global climate change. There is no excuse for lack of preparedness from our leaders. The way migration is 'managed' today is a moral, social and economic failure – lives are being needlessly wasted daily. It is time to begin the conversation about how we address this; it is time for a global collaborative solution both to climate change and to migration, via a diverse and inclusive pool of global decision makers.

At some point between now and 2030, we will have to accept that limiting temperature rise to 1.5°C isn't feasible through our mitigation efforts. We will have to decide whether to use solar reflection or focus on adapting to more dangerous temperatures. This, at a time when half of the world's 2.2 billion children are already at 'extremely high risk' of the impacts of climate change.[3]

Climate change is *everything* change, because climate is the fabric on which we weave our lives. It determines where human habitation is viable and how we live there, the timing of seasons and what can be grown, where the rain falls, how hot it is, the shape of coastlines and extent of landscapes, the routing of ocean currents and the severity of storms. Every one of us will experience this profound existential change over the coming decades – a dislocation in our relationship with the environment that generated our culture, our society, our own lives. Climate change is not smooth and predictable but erratic over time, and we will experience our dislocation in jolts. In 2021, a heatwave along the Pacific coast of North America cooked more than 1 billion seashore creatures, many in their shells; fruit was cooked on the trees; crops and buildings were burned; and hundreds of people died. Extreme conditions will hit the globe with increasing frequency, pushing the sudden exodus of populations.

Today, renewables are adding to, not replacing, fossil fuels, thus temperatures are rising. Almost one-third of the carbon emissions ever produced in the entire history of humanity have been since the release of Al Gore's *An Inconvenient Truth* in 2006, so I'm not convinced greater awareness of the problem is what's lacking. By 2050, when more than 1 billion people are on the move, we need to have figured out a way of managing this to the benefit of all of us.

We are facing existential threats: almost incomprehensible in scale, and yet real and soon. The planetary nature of humanity's crisis can feel overwhelming, and our situation hopeless. It is not.

Planning for this upheaval is about building the resilience of our species in the face of an uncertain future. Uncertainty stops us planning, preparing, and even looking too far into the future. There is a reluctance to go beyond an extrapolation of the present to imagine familiar landscapes transformed. Nevertheless, hard questions must be confronted about what will be possible in a hotter world. This means actively provoking people to raise their heads from managing their day-to-day priorities and getting them to imagine themselves twenty, thirty, forty years into the future.[4] Keep an open mind as you think through all of the ways we might reach it, setting aside knee-jerk biases towards or against the different possibilities. Consider this future world with empathy for your elderly self – what sort of society will support you? Will it be young, vibrant and hopeful, or will you be fearful and uncared for?

My point is that we can build resilience into our social systems – and our ecosystems – to withstand the stresses and shocks of climate change and disasters. But we need to own our future. This means agreement on a future vision, on what outcome would retain the most value in terms of our cherished cultures, our lives of comfort and security, and our natural world.

We are not impotent bystanders. But today we lack a coherent plan; we are simply experiencing our world heating up, and reacting to each new shock – each drought, each typhoon, each blazing forest, each heaving boat of migrants – with a new patch-up. We must take control of our future, and that means making a plan to protect the well-being of all humans, rich and poor, from every continent, as we enter the challenging environment of the coming decades. This means having the courage to envision a different way of being a human: in effect, unsticking people from their fixed abodes and setting them free to roam, free to seek the safe places.

During the Covid pandemic, we transformed our understanding of what is normal and what is socially possible. Who would have believed that so many of us would have voluntarily restricted our movements to within metres of our homes? It is, I think, easier to imagine the opposite: that many of us will move thousands of kilometres from home.

People will move in their millions – right now, we have a chance to make it work. This could be a planned, managed, peaceful transition to a safer, fairer world. With international cooperation and regulation, we could and should make the Earth liveable.

That has to be worth trying. So let's begin.

Manifesto

1. People relocating is a natural human behaviour; migration is a successful survival adaptation.
2. We need to redirect the productive capacity of society to address climate change and the looming demographic crisis.
3. We must ensure a safe, fair process for migration, overseen by a global agency with powers to police it.
4. Migration is an economic not a security issue. It drives economic growth and reduces poverty.
5. Rich countries and poor countries must invest in alliances that increase training and education, and climate resilience.
6. Decarbonizing our economies must be done urgently and globally, including through taxation and incentives.
7. Ice melt and coral reef loss are already dangerously accelerating: solar reflectivity, such as cloud brightening, should be deployed without delay, and other technologies to reduce temperatures should be explored.
8. We must work urgently to reverse the destruction of ecosystems and restore biodiversity to build resilience and protect natural systems.

Further Reading

For a deeper dive into the issues I cover in *Nomad Century*, here is a selection of jumping off points – check my website, *WanderingGaia. com*, for the latest suggestions. My own books *Adventures in the Anthropocene* and *Transcendence* also contain much relevant material and context.

Akala, *Natives: Race and class in the ruins of empire* (Hodder & Stoughton, 2021)

Abhijit V. Banerjee and Esther Duflo, *Good Economics for Hard Times* (Allen Lane, 2019)

Paul Behrens, *The Best of Times, the Worst of Times: Futures from the frontiers of climate science* (Indigo Press, 2021)

Mike Berners-Lee, *There is No Planet B* (Cambridge University Press, 2019)

Sally Hayden, *My Fourth Time, We Drowned: Seeking refuge on the world's deadliest migration route* (Fourth Estate, 2022)

Eric Holthaus, *The Future Earth: A radical vision for what's possible in the age of warming* (HarperOne, 2020)

Rowan Hooper, *How to Spend a Trillion Dollars* (Profile Books, 2021)

Elizabeth Kolbert, *Under a White Sky* (Vintage, 2022)

J. Krause and T. Trappe, *A Short History of Humanity: How migration made us who we are* (W. H. Allen, 2021)

Felix Marquardt, *New Nomads* (Simon & Schuster, 2021)

John Pickrell, *Flames of Extinction: The race to save Australia's threatened wildlife* (Island Press, 2021)

J. Purdy, *This Land is Our Land: The struggle for a new commonwealth* (Princeton University Press, 2020)

Kim Stanley Robinson, *The Ministry for the Future* (Orbit, 2020)

Doug Saunders, *Arrival City* (Vintage, 2012)

Laurence Smith, *The New North: The World in 2050* (Dutton Books, 2010)

—— *Rivers of Power: How a natural force raised kingdoms, destroyed civilizations, and shapes our world* (Little, Brown Spark, 2021)

Carolyn Steel, *Sitopia: How food can save the world* (Chatto & Windus, 2020)

Acknowledgements

This book is a collection of several years' research that would have been impossible without the assistance and kindness of so many people across the world. I thank all those who have helped me understand the lives of others and shared their stories with me. I've also relied on the knowledge and wisdom of experts who took their time to talk to me and explain their research, for which I am very grateful. Particular thanks to Margaret Young, Duncan Graham-Rowe, Mun-Keat Looi, Oli Franklin-Wallis, Deborah Cohen, Richard Betts, Laurence Smith, Doug Saunders, Hannah Ritchie, Max Roser, David King, David Keith, Jesse Reynolds, Chris Smaje, Neil Adger, Mariana Mazzucato, Ken Caldeira, Alex Randall, Stijn Hoorens, Fatih Birol, and teams at Gavi, UNICEF and UNHCR.

This book wouldn't exist without the support and encouragement of my brilliant agent and friend Patrick Walsh, and the wonderful John Ash and Margaret Halton at PEW Literary. Huge thanks also to the excellent teams at Allen Lane and Flatiron Books, especially to my fabulous editors, the brilliant Laura Stickney and Lee Oglesby, who provided essential advice and helped shape the book. Thanks also to Sam Fulton, Jane Birdsell and Richard Duguid.

I wrote it during the Covid pandemic with its lockdowns, homeschooling, and other ruinous obstacles, an ordeal only made possible through the love and friendship of family and friends, including Jolyon Goddard, Rowan Hooper, Olive Heffernan, Sara Abdulla, John Witfield, Charlotte and Henry Nicholls, my Sisters of the Pen: Jo Marchant and Emma Young, my parents Ivan and Gina, my brother David, and my raisons d'être Nick, Kipp and Juno. Thanks for enduring the moaning, next time it'll be better . . .

Notes

INTRODUCTION

1. United Nations (Department of Economic and Social Affairs), *World Population Ageing 2019* (ST/ESA/SER. A/444) (2020).
2. https://www.climate.gov/news-features/blogs/beyond-data/2021-us-billion-dollar-weather-and-climate-disasters-historical

1. The Storm

1. Two-thirds of those asked in a global poll of more than 1 million people. United Nations Development Programme: https://www.undp.org/publications/g20-peoples-climate-vote-2021
2. Alan M. Haywood, Harry J. Dowsett and Aisling M. Dolan, 'Integrating geological archives and climate models for the mid-Pliocene warm period', *Nature Communications* 7:1 (2016), pp. 1–14.
3. The immediate greenhouse effect of this CO_2 release was entirely offset by vast amounts of sulphur that were also emitted, forming a reflective barrier to sunlight. Global temperatures plummeted, and the lack of sunlight and changes in ocean circulation devastated Earth's plant life and led to mass extinctions.
4. Of global heating above pre-industrial levels.
5. J. M. Murphy, G. R. Harris, D. M. H. Sexton, *et al.*, 'UKCP18 land projections: Science report' (UK Met Office, 2018).
6. Thomas Slater, Isobel R. Lawrence, Inès N. Otosaka, *et al.*, 'Earth's ice imbalance', *The Cryosphere* 15:1 (2021), pp. 233–46.
7. CIRES, *The Threat from Thwaites: The retreat of Antarctica's riskiest glacier* (2021). Available at: https://cires.colorado.edu/news/threat-thwaites-retreat-antarctica's-riskiest-glacier and https://www.youtube.com/watch?v=uBbgWsR4-aw

8. N. Boers and M. Rypdal, 'Critical slowing down suggests that the western Greenland Ice Sheet is close to a tipping point', *Proceedings of the National Academy of Sciences* 118:21 (2021), p.e2024192118.

9. In the Miocene, as CO_2 increased from just below today's levels up to about 500 ppm, Antarctica shed what would amount to 30–80 per cent of the modern ice sheet. The Antarctic ice sheet may be more vulnerable today to rapid retreat and disintegration than at any time in its entire 34 million-year history.

10. N. Burls and A. Fedorov, 'Wetter subtropics in a warmer world: Contrasting past and future hydrological cycles', *Proceedings of the National Academy of Sciences* 114:49 (2017), pp.12888–93.

11. Charles Geisler and Ben Currens, 'Impediments to inland resettlement under conditions of accelerated sea level rise', Land Use Policy 66 (2017), p. 322. [DOI: 10.1016/j.landusepol.2017.03.029]

12. Samantha Bova, Yair Rosenthal, Zhengyu Liu, *et al.*, 'Seasonal origin of the thermal maxima at the Holocene and the last interglacial', *Nature* 589:7843 (2021), pp. 548–53.

13. Bear in mind that humans evolved in the Pleistocene, a time of punishing ice ages, and our species has spent most of its evolutionary history struggling to survive these harsh conditions. This itself is unusual – for almost all of the Earth's history, the planet was a much warmer place than it is today, with much higher CO_2 levels, courtesy of volcanoes.

14. Paul J. Durack, Susan E. Wijffels and Richard J. Matear, 'Ocean salinities reveal strong global water cycle intensification during 1950 to 2000', *Science* 336:6080 (2012), pp. 455–8.

15. Aslak Grinsted and Jens Hesselbjerg Christensen, 'The transient sensitivity of sea level rise', *Ocean Science* 17:1 (2021), pp. 181–6.

2. The Four Horsemen of the Anthropocene

1. 'Pyrocene' was coined by the US environmental historian Stephen J. Pyne.

2. In August, the state experienced its first gigafire, when more than 1 million acres burned.

3. Adam M. Young, Philip E. Higuera, Paul A. Duffy and Feng Sheng Hu, 'Climatic thresholds shape northern high-latitude fire regimes and imply vulnerability to future climate change', *Ecography* 40:5 (2017), pp. 606–17.

4. D. Bowman, G. Williamson, J. Abatzoglou, *et al.*, 'Human exposure and sensitivity to globally extreme wildfire events', *Nature Ecology & Evolution* 1 (2017), e0058.

5. Merritt R. Turetsky, Brian Benscoter, Susan Page, *et al.*, 'Global vulnerability of peatlands to fire and carbon loss', *Nature Geoscience* 8:1 (2015), pp. 11–14.

6. Saul Elbein, 'Wildfires threaten California communities on new financial front', *The Hill*, 8 April 2021. Available at: https://thehill.com/policy/equilibrium-sustainability/566360-wildfires-threaten-california-communities-on-new-financial

7. A. McKay, 'Just had my home insurance cancelled because Southern California is at too high risk now for fire and floods. This shit is real and happening right now. #endfossilfuels #dontlookup', Twitter, 14 January 2022: https://twitter.com/ghostpanther/status/1482064740482359297

8. State of California, 'Commissioner Lara protects more than 25,000 policyholders affected by Beckwourth Complex Fire and lava fire from policy non-renewal for one year', CA Department of Insurance (n.d.).

9. Intergovernmental Panel on Climate Change, '5: Changing ocean, marine ecosystems, and dependent communities' in *IPCC Special Report on the Ocean and Cryosphere in a Changing Climate* (2019). Available at: https://www.ipcc.ch/srocc/

10. Chi Xu, Timothy A. Kohler, Timothy M. Lenton, *et al.*, 'Future of the human climate niche', *Proceedings of the National Academy of Sciences* 117:21 (2020), pp. 11350–55.

11. The US National Weather Service defines a 'heat index' – combining heat and humidity into a 'feels-like' temperature – of 40.6°C as 'dangerous'.

12. Eun-Soon Im, Jeremy S. Pal and Elfatih A. B. Eltahir, 'Deadly heatwaves projected in the densely populated agricultural regions of South Asia', *Science Advances* 3:8 (2017), e1603322.

13. Ibid.

14. Tamma A. Carleton, Amir Jina, Michael T. Delgado, *et al.*, *Valuing the Global Mortality Consequences of Climate Change Accounting for Adaptation Costs and Benefits*, National Bureau of Economic Research, working paper 27599 (2020).

15. Yan Meng and Long Jia, 'Global warming causes sinkhole collapse: Case study in Florida, USA', *Natural Hazards and Earth System Sciences Discussions* (2018), pp. 1–8.

16. Because hotter air is less dense, meaning air pressure under the wings is insufficient for take-off. Compounding the problem, less oxygen

flows through the engine, reducing its ability to create thrust. And extreme heat can create air turbulence and also buckle and crack the runway.

17. F. Bassetti, 'Environmental migrants: Up to 1 billion by 2050', *Foresight*, 3 August 2021.

18. Zhao Liu, Bruce Anderson, Kai Yan, *et al.*, 'Global and regional changes in exposure to extreme heat and the relative contributions of climate and population change', *Scientific Reports* 7:1 (2017), pp. 1–9.

19. Ethan D. Coffel, Radley M. Horton and Alex De Sherbinin, 'Temperature and humidity based projections of a rapid rise in global heat stress exposure during the 21st century', *Environmental Research Letters* 13:1 (2017), e014001.

20. Yuming Guo, Antonio Gasparrini, Shanshan Li, *et al.*, 'Quantifying excess deaths related to heatwaves under climate change scenarios: A multicountry time series modelling study', *PLoS Medicine* 15:7 (2018), e1002629.

21. Suchul Kang and Elfatih A. B. Eltahir, 'North China Plain threatened by deadly heatwaves due to climate change and irrigation', *Nature Communications* 9:1 (2018), pp. 1–9.

22. S. Mufson, 'Facing unbearable heat, Qatar has begun to air-condition the outdoors', *Washington Post*, 16 October 2019.

23. E. Team, 'This air-conditioned suit lets you work outside on even the hottest days', eeDesignIt.com, 13 December 2017.

24. M. Nguyen, 'To beat the heat, Vietnam rice farmers resort to planting at night', Reuters, 25 June 2020.

25. Nick Watts, Markus Amann, Nigel Arnell, *et al.*, 'The 2019 report of The Lancet Countdown on health and climate change: ensuring that the health of a child born today is not defined by a changing climate', *The Lancet* 394: 10211 (2019), pp. 1836–78.

26. O. Milman and A. Chang, 'How heat is radically altering Americans' lives before they're even born – video', *The Guardian*, 16 February 2021.

27. Sara McElroy, Sindana Ilango, Anna Dimitrova, *et al.*, 'Extreme heat, preterm birth, and stillbirth: A global analysis across 14 lower-middle income countries', *Environment International* 158 (2022), e106902.

28. Marco Springmann, Daniel Mason-D'Croz, Sherman Robinson, *et al.*, 'Global and regional health effects of future food production under climate change: A modelling study', *The Lancet* 387:10031 (2016), pp. 1937–46.

29. IPCC, *Climate Change 2022: Impacts, adaptation, and vulnerability. Contribution of Working Group II to the Sixth Assessment Report*

of the Intergovernmental Panel on Climate Change, ed. H-O. Pörtner, D. C. Roberts, M. Tignor, *et al.* (Cambridge University Press, in press).

30. 'Report: Flooded future: Global vulnerability to sea level rise worse than previously understood', *Climate Central*, 29 October 2019.

31. Svetlana Jevrejeva, Luke P. Jackson, Riccardo E. M. Riva, *et al.*, 'Coastal sea level rise with warming above 2°C', *Proceedings of the National Academy of Sciences* 113:47 (2016), pp. 13342–7.

32. K. Mohammed, A. K. Islam, G. M. Islam, *et al.*, 'Future floods in Bangladesh under 1.5°C, 2°C, and 4°C global warming scenarios', *Journal of Hydrologic Engineering* 23:12 (2018), e04018050; https://doi.org/10.1061/(asce)he.1943-5584.0001705

33. Stein Emil Vollset, Emily Goren, Chun-Wei Yuan, *et al.*, 'Fertility, mortality, migration, and population scenarios for 195 countries and territories from 2017 to 2100: A forecasting analysis for the Global Burden of Disease Study', *The Lancet* 396:10258 (2020), pp. 1285–1306.

3. Leaving Home

1. Scott R. McWilliams and William H. Karasov, 'Migration takes guts' in *Birds of Two Worlds: The ecology and evolution of migration* (Smithsonian Institution Press, Washington, DC, 2005), pp. 67–78.

2. Dominique Maillet and Jean-Michel Weber, 'Performance-enhancing role of dietary fatty acids in a long-distance migrant shorebird: the semi-palmated sandpiper', *Journal of Experimental Biology* 209:14 (2006), pp. 2686–95.

3. Scientists have identified what's been described as an 'explorer gene' (gene G4GR) in our DNA, which may have helped our ancestors adapt migratory behaviours when following moving herds of herbivores, with changing water availability; and helped diversify our gene pool beyond the population found in one location.

4. We know that farming spread widely, but until recently, we've not known how – whether it was the idea that migrated, or whether it was the people themselves. Genetic analysis indicates that both took place. Within the Fertile Crescent, farming knowledge and tools were developed and exchanged between populations. Small groups of these farmers then migrated from Anatolia and the Levant around 9,000 to 7,000 years ago, bringing their new expertise in seed-collecting and sowing, brewing and animal husbandry to Europe and East Africa – one-third of Somali DNA comes from the Levant population.

5. Elizabeth Gallagher, Stephen Shennan and Mark G. Thomas, 'Food income and the evolution of forager mobility', *Scientific Reports* 9:1 (2019), pp. 1–10.

6. David Kaniewski, Joël Guiot and Elise Van Campo, 'Drought and societal collapse 3200 years ago in the Eastern Mediterranean: A review', *Wiley Interdisciplinary Reviews: Climate Change* 6:4 (2015), pp. 369–82.

7. S. Solomon, 'The future is mixed-race and that's a good thing for humanity', *Aeon*, 19 February 2022.

8. D. Varinsky, 'Cities are becoming more powerful than countries', *Business Insider*, 19 August 2016.

4. Bordering on Insanity

1. S. Loarie, P. Duffy, H. Hamilton, *et al.*, 'The velocity of climate change', *Nature* 462 (2009), pp. 1052–5.

2. Jonathan Woetzel, Anu Madgavkar and Khaled Rifai, *People on the Move: Global migration's impact and opportunity* (McKinsey Global Institute, 2016).

3. A. Gaskell, 'The economic case for open borders', *Forbes*, 21 January 2021.

4. I. Dias, 'One man's quest to crack the modern anti-immigration movement by unsealing its architect's papers', *Mother Jones*, 30 March 2021.

5. David A. Bell, *The Cult of the Nation in France* (Harvard University Press, Cambridge, MA, 2001).

6. M. Nagdy and M. Roser, 'Civil wars', *Our World in Data*. Available at: https://ourworldindata.org/civil-wars

7. Those former colonies that had complex bureaucracies, such as India, tended to become relatively stable nation states. Whereas those without, where colonial rulers had only extracted resources, such as the former Belgian Congo, did not go on to become stable democracies.

8. M. Salvini, 'A Tripoli coi ragazzi della Caprera, che difendono I Mari e la Nostra Sicurezza: Onore!', Twitter, 25 June 2018: pic.twitter.com/wgbgctvut7

9. Chi Xu, Timothy A. Kohler, Timothy M. Lenton, *et al.*, 'Future of the human climate niche', *Proceedings of the National Academy of Sciences* 117:21 (2020), pp. 11350–55.

10. This new global precedent actually came out of a case in which the committee sided with a state's *refusal* to provide asylum. Ioane Teitiota, originally from the Pacific island nation of Kiribati, migrated with his family to New Zealand, where he applied for refugee status after his visa expired in 2010.

Teitiota argued that, because his home island of South Tarawa is expected to become uninhabitable in the next ten to fifteen years, his family's lives were at risk. New Zealand rejected Teitiota's claim for asylum, and the UN upheld the decision, because Kiribati was not yet uninhabitable, so this 'could allow for intervening acts by the republic of Kiribati, with the assistance of the international community, to take affirmative measures to protect and, where necessary, relocate its population'. However, the committee also ruled that the climate crisis could 'expose individuals to a violation of their rights' which would, in turn, prohibit states under international law from sending refugees back to their home countries. The committee pointed to articles 6 and 8 of the International Covenant on Civil and Political Rights, which secures a person's right to life.

11. Climate change isn't the only environmental risk to life recognized by courts recently. In 2021, a French court made legal history by taking into account the environmental conditions in a migrant's country of origin in his appeal against deportation back to Bangladesh. The man was allowed to remain in France because the dangerous levels of air pollution in his birth country meant it was unsafe to send him back.

5. Wealth of Migrants

1. B. D. Caplan, Z. Weiner and M. Cagle, *Open Borders: The science and ethics of immigration* (St Martin's Press, 2019).

2. Paul Almeida, Anupama Phene and Sali Li, 'The influence of ethnic community knowledge on Indian inventor innovativeness', *Organization Science* 26:1 (2015), pp. 198–217.

3. Caroline Theoharides, 'Manila to Malaysia, Quezon to Qatar: International migration and its effects on origin-country human capital', *Journal of Human Resources* 53:4 (2018) pp. 1022–49.

4. John Gibson and David McKenzie, 'Eight questions about brain drain', *Journal of Economic Perspectives* 25:3 (2011) pp. 107–28.

5. https://www.oecd.org/dev/development-posts-Global-Skill-Partnerships-A-proposal-for-technical-training-in-a-mobile-world.htm

6. Jonathan Woetzel, Anu Madgavkar and Khaled Rifai, *People on the Move: Global migration's impact and opportunity* (McKinsey Global Institute, 2016).

7. Jonathan Woetzel, Anu Madgavkar and Khaled Rifai, *People on the move: Global migration's impact and opportunity* (McKinsey Global Institute, 2016).

8. Average incomes were 20 per cent higher in counties with median immigrant inflows relative to counties with no immigrant inflows, the proportion of people living in poverty was 3 percentage points lower, the unemployment rate was 3 percentage points lower, the urbanization rate was 31 percentage points higher, and education attainment was higher as well. Sandra Sequeira, Nathan Nunn and Nancy Qian, *Migrants and the making of America: The short- and long-run effects of immigration during the age of mass migration*, National Bureau of Economic Research, working paper 23289 (2017).

9. G. Peri, 'Immigration, labor markets, and productivity', *Cato Journal* 32:1 (2012), pp. 35–53.

10. As has been eloquently explained by the Nobel Prize-winning economists Abhijit V. Banerjee and Esther Duflo, for example in their book *Good Economics for Hard Times*.

11. Andrew Nash, 'National population projections: 2016-based statistical bulletin', Office for National Statistics (ONS), October 2015.

12. Baby boomers have consolidated their position through soaring property prices, inheritance and the prevalence of final salary pension schemes. C. Canocchi, 'One in five baby boomers are now millionaires as their wealth doubles', *This is Money*, 15 January 2019.

13. J. P. Aurambout, M. Schiavina, M. Melchiori, *et al.*, *Shrinking Cities* (European Commission, 2021; JRC126011).

14. Michael A. Clemens, 'Economics and emigration: Trillion-dollar bills on the sidewalk?', *Journal of Economic Perspectives* 25:3 (2011), pp. 83–106.

15. International Labour Conference, 92nd Session, *Towards a Fair Deal for Migrant Workers in the Global Economy: Report VI* (2004).

16. 'Immigration has been and continues to be an important driver of Australian growth', from Dr Stephen Kennedy's speech, 'Australia's response to the global financial crisis'. Available at: https://treasury.gov.au/speech/australias-response-to-the-global-financial-crisis

17. Jose-Louis Cruz and Esteban Rossi-Hansberg, *The Economic Geography of Global Warming*, National Bureau of Economic Research, working paper 28466 (2021).

18. Hein De Haas, 'International migration, remittances and development: Myths and facts', *Third World Quarterly* 26:8 (2005), pp. 1269–84.

19. Paul Clist and Gabriele Restelli, 'Development aid and international migration to Italy: Does aid reduce irregular flows?', *World Economy* 44:5 (2021), pp. 1281–1311.

6. New Cosmopolitans

1. E. McConnell, ''27 people drowned and I laughed', *Kent Online*, 30 November 2021.
2. D. Bahar, P. Choudhury and B. Glennon, 'The day that America lost $100 billion because of an immigration visa ban', *Brookings*, 20 October 2020.
3. Tim Wadsworth, 'Is immigration responsible for the crime drop? An assessment of the influence of immigration on changes in violent crime between 1990 and 2000', *Social Science Quarterly* 91:2 (2010), pp. 531–53.
4. Eurostat 2021, 'Asylum applicants by type of applicant, citizenship, age and sex – monthly data (rounded)'.
5. Vera Messing and Bence Ságvári, 'Are anti-immigrant attitudes the Holy Grail of populists? A comparative analysis of attitudes towards immigrants, values, and political populism in Europe', *Intersections: East European Journal of Society and Politics* 7:2 (2021), pp. 100–127.
6. B. Stokes, 'How countries around the world view national identity', *Pew Research Center's Global Attitudes Project*, 30 May 2020.
7. Vincenzo Bove and Tobias Böhmelt, 'Does immigration induce terrorism?', *Journal of Politics* 78:2 (2016), pp. 572–88.
8. One report by the US Bureau of Labor Statistics projects that 'people of colour will become a majority of the American working class in 2032'. Valerie Wilson, 'People of color will be a majority of the American working class in 2032', *Economic Policy Institute* 9 (2016).
9. The majority of migrants are far more interested in mobility than voting rights, according to surveys.
10. The *hukou* system has bestowed far greater benefits on urban households compared to rural households, exacerbating inequality. *Hukou* reforms are ongoing . . .
11. D. Held, *Democracy and the Global Order: From the modern state to cosmopolitan governance* (Polity Press, 1995).
12. David Miller, *On Nationality* (Clarendon Press, 1995).
13. Anna Marie Trester, *Bienvenidos a Costa Rica, la tierra de la pura vida: A Study of the Expression 'pura vida' in the Spanish of Costa Rica* (2003), pp. 61–9. Available at: https://georgetown.academia.edu/AnnaMarieTrester

7. Haven Earth

1. Double this area to include street infrastructure and you could still fit it all into France.

2. Chi Xu, Timothy A. Kohler, Timothy M. Lenton, *et al.*, 'Future of the human climate niche', *Proceedings of the National Academy of Sciences* 117:21 (2020), pp. 11350–55.

3. J. Kevin Summers, Linda C. Harwell, Kyle D. Buck, *et al.*, *Development of a Climate Resilience Screening Index (CRSI): An assessment of resilience to acute meteorological events and selected natural hazards* (US Environmental Protection Agency, Washington, DC, 2017).

4. Camilo Mora, Abby G. Frazier, Ryan J. Longman, *et al.*, 'The projected timing of climate departure from recent variability', *Nature* 502:7470 (2013), pp. 183–7.

5. C. Welch, 'Climate change has finally caught up to this Alaska village', *National Geographic* 22 (2019).

6. In August 2021, the Danish Meteorological Institute reported temperatures of more than 20°C – more than twice the normal average summer temperature – in northern Greenland.

7. Signe Normand, Christophe Randin, Ralf Ohlemüller, *et al.*, 'A greener Greenland? Climatic potential and long-term constraints on future expansions of trees and shrubs', *Philosophical Transactions of the Royal Society B: Biological Sciences* 368:1624 (2013), e20120479.

8. Svante Arrhenius, 'XXXI. On the influence of carbonic acid in the air upon the temperature of the ground', *London, Edinburgh, and Dublin Philosophical Magazine and Journal of Science* 41:251 (1896), pp. 237–76.

9. Ove Hoegh-Guldberg, Marco Bindi and Myles Allen, 'Chapter 3: Impacts of 1.5°C global warming on natural and human systems 2' in *Global warming of 1.5°C: An IPCC Special Report* (2018).

10. J. Garthwaite, 'Climate change has worsened global economic inequality', *Stanford University Earth Matters Magazine* (2019).

11. P. T. Finnsson and A. Finnsson, 'The Nordic region could reap the benefits of a warmer climate', *NordForsk* 4, 13 September 2014; available at: https://partner.sciencenorway.no/agriculture-climate-change-farming/the-nordic-region-could-reap-the-benefits-of-a-warmer-climate/1406934

12. The greening of the tundra, as hardy birch trees take root, further accelerates the warming process, as the birch improves the soil and warms it with microbial activity, melting the permafrost and releasing methane – a greenhouse gas eighty-five times more powerful than carbon dioxide in its warming effects, over a shorter timeframe.

13. Daniela Jacob, Lola Kotova, Claas Teichmann, *et al.*, 'Climate impacts in Europe under +1.5 C global warming', *Earth's Future* 6:2 (2018), pp. 264–85.

14. K. El-Assal, 'Canada breaks all-time immigration record by landing 401,000 immigrants in 2021', *Canada Immigration News*, 23 January 2022; https://www.cicnews.com/2021/12/canada-breaks-all-time-immigration-record-by-landing-401000-immigrants-in-2021-1220461.html#gs.u4894i

15. Marshall Burke, Solomon M. Hsiang and Edward Miguel, 'Global non-linear effect of temperature on economic production', *Nature* 527:7577 (2015), pp. 235–9.

16. Elena Parfenova, Nadezhda Tchebakova and Amber Soja, 'Assessing landscape potential for human sustainability and "attractiveness" across Asian Russia in a warmer 21st century', *Environmental Research Letters* 14:6 (2019), e065004.

17. Jan Hjort, Dmitry Streletskiy, Guy Doré, *et al.*, 'Impacts of permafrost degradation on infrastructure', *Nature Reviews Earth & Environment* 3:1 (2022), pp. 24–38.

18. 'A lot of Arctic infrastructure is threatened by rising temperatures', *The Economist*, 15 January 2022.

19. Russia's longstanding xenophobia is entrenched and wide-ranging, including negativity to Jews, Roma, Chinese and Vietnamese.

20. 'Agreement of Coexistence … which stipulates that the resident is explicitly, freely, and voluntarily consenting to the governance structures, rulemaking systems, and authority of Próspera'.

21. Julia Carrie Wong, 'Seasteading: Tech leaders' plans for floating city trouble French Polynesians', *The Guardian*, 2 January 2017.

22. Costas Meghir, Ahmed Mushfiq Mobarak, Corina D. Mommaerts and Melanie Morten, *Migration and Informal Insurance: Evidence from a randomized controlled trial and a structural model*, National Bureau of Economic Research, working paper 26082 (2019).

23. Ahmed Mushfiq Mobarak, 'Can a bus ticket prevent seasonal hunger?', *Yale Insights* 18 (2018).

24. B. Lyte, 'Remote workers are flocking to Hawaii: But is that good for the islands?, *The Guardian*, 26 January 2021.

25. Tatyana Deryugina, Laura Kawano and Steven Levitt, 'The economic impact of Hurricane Katrina on its victims: Evidence from individual tax returns', *American Economic Journal: Applied Economics* 10:2 (2018), pp. 202–33.

8. Migrant Homes

1. Allen J. Scott, 'World Development Report 2009: Reshaping economic geography', *Journal of Economic Geography* 9:4 (2009), pp. 583–6.

2. Joaquin Arango, Ramon Mahia, David Moya Malapeira and Elena Sanchez-Montijano, 'Introduction: Immigration and asylum, at the center of the political arena', *Anuario CIDOB de la Inmigracion* (2018), pp. 14–26.

3. Phillip Connor, 'A majority of Europeans favor taking in refugees, but most disapprove of EU's handling of the issue', *Pew Research Center*, 19 September 2018.

4. Joaquín Arango, *Exceptional in Europe? Spain's experience with immigration and integration*, Migration Policy Institute (March 2013).

5. Live tables on land use, available at: https://www.gov.uk/government/statistical-data-sets/live-tables-on-land-use

6. N. Gabobe, 'Living together: It's time for zoning codes to stop regulating family type', Sightline Institute, 28 February 2020.

7. Araveno has since put all the architectural designs online to be freely used, and other cities are using them.

8. 'The world's first affordable 3D printed village pops-up [*sic*] in Mexico', *The Spaces*, 13 December 2019.

9. Anthropocene Habitats

1. *Urbanisation and Climate Change Risks: Environmental risk outlook 2021* (Verisk Maplecroft); available at: https://www.maplecroft.com/insights/analysis/asian-cities-in-eye-of-environmental-storm-global-ranking/

2. Sea-level increase for coastal cities is 7.8–9.9mm annually, compared to the global average of 2.5mm a year for the past two decades.

3. Mark Fischetti, 'Sea level could rise 5 feet in New York City by 2100', *Scientific American*, 1 June 2013.

4. S. Fratzke and B. Salant, *Moving beyond Root Causes: The complicated relationship between development and migration* (Migration Policy Institute, January 2018).

5. https://maldivesfloatingcity.com

6. O. Wainwright, 'A £300 monsoon-busting home: The Bangladeshi architect fighting extreme weather', *The Guardian*, 16 November 2021.

7. *Chicago Sustainable Development Policy, 2017* (City of Chicago).

8. Heat Island Group, Berkley Lab, 'Cool roofs'; available at: https://heatisland.lbl.gov/coolscience/cool-roofs

9. The scientists who developed the paint, which is made with barium sulphate, estimate that painting 1 per cent of the Earth's surface with this paint – 'perhaps an area where no people live that is covered in

rocks' – could significantly offset global heating; https://www.bbc.co.uk/news/science-environment-56749105

10. Jonathon Laski and Victoria Burrows, *From Thousands to Billions: Coordinated action towards 100 per cent net zero carbon buildings by 2050*, World Green Building Council (2017).

11. '19 global cities commit to make new buildings "net-zero carbon" by 2030', *C40 Cities*, 15 October 2021.

12. M. De Socio, 'The US city that has raised $100m to climate-proof its buildings', *The Guardian*, 19 August 2021.

13. One Norwegian town, Rjukan, built in the shadow of mountains around a hydropower complex, has geoengineered a way around the problem, by erecting large rotatable mirrors that beam precious sunlight into the town, extending the day.

14. Alex Nowrasteh and Andrew C. Forrester, *Immigrants Recognize American Greatness: Immigrants and their descendants are patriotic and trust America's governing institutions*, Cato Institute, Immigration research and policy brief 10 (2019).

10. Food

1. L. Hengel, 'Famine alert: How WFP is tackling this other deadly pandemic', UN World Food Programme website, 29 March 2021.

2. Zhu Zhongming, Lu Linong, Zhang Wangqiang and Liu Wei, 'Impact of climate change on crops adaptation and strategies to tackle its outcome: A review', *Plants* 8:2 (2019), p. 34.

3. Matti Kummu, Matias Heino, Maija Taka, *et al.*, 'Climate change risks pushing one-third of global food production outside the safe climatic space', *One Earth* 4:5 (2021), pp. 720–29.

4. Chuang Zhao, Bing Liu, Shilong Piao, *et al.*, 'Temperature increase reduces global yields of major crops in four independent estimates', *Proceedings of the National Academy of Sciences* 114:35 (2017), pp. 9326–31.

5. Andrew S. Brierley and Michael J. Kingsford, 'Impacts of climate change on marine organisms and ecosystems', *Current Biology* 19:14 (2009), pp. R602–14.

6. B. Byrne, *2020 State of the Industry Report: Cultivated meat*, Good Food Institute.

7. Keri Szejda, Christopher J. Bryant and Tessa Urbanovich, 'US and UK consumer adoption of cultivated meat: A segmentation study', *Foods* 10:5 (2021), p. 1050.

8. Björn Witte, Przemek Obloj, Sedef Koktenturk, *et al.*, 'Food for thought: The protein transformation', *Industrial Biotechnology* (2021).

9. Myron King, Daniel Altdorff, Pengfei Li, *et al.*, 'Northward shift of the agricultural climate zone under 21st-century global climate change', *Scientific Reports* 8:1 (2018), pp. 1–10.

10. D. Singer, 'The drones watching over cattle where cowboys cannot reach', BBC website, n.d.

11. Ward Anseeuw and Giulia Maria Baldinelli, *Uneven Ground: Research findings from the Land Inequality Initiative* (International Fund for Agricultural Development, 2020).

12. 'Six ways indigenous peoples are helping the world achieve zero hunger', Food and Agriculture Organization of the UN. Available at: https://www.fao.org/indigenous-peoples/news-article/en/c/1029002/

11. Power, Water, Stuff

1. Floating solar is at least 10 per cent costlier than land-based installations, but panels on water don't get as dusty, perform better because they are cooler and, best of all, floating solar requires less new land to be used for solar or flooded for hydro.

2. In China, the 22,500 megawatt Three Gorges Dam created the world's longest lake from the Yangtse River and in the process evicted 1.3 million people, drowned thirteen cities and 1,500 towns and villages, and destroyed numerous ecological and cultural sites.

3. In its 2021 road map to net zero by 2050, the IEA factored in an extraordinary growth in nuclear capacity, reaching 30 gigawatt hours per year by 2030, which is five times the rate of the past decade, so that by 2050 global installed capacity would be double today's. Given the state of investment in the sector in recent decades and the shutdowns many countries plan, this seems ambitious and would require extremely rapid deployment.

4. One innovative plan, described as 'high-intensity gentle-slope' hydropower, involves using a mineral fluid with twice the density of water for underground pumped hydro, so that common hills with small drops can be used to store energy.

5. But in general, and certainly in the EU, using renewable energy to create hydrogen must be weighed against the many other jobs that renewables have to do in our green-energy economy, and producing electricity to power our world is a more efficient way of using wind and solar power

than using it to split water to make hydrogen. In other words, prioritize electrification and don't get distracted by hydrogen projects, particularly if they involve heating homes or fuelling vehicles, which can be done far more efficiently with heat pumps and batteries respectively.

6. Shipping today is a particularly filthy business because of the low-grade fossil fuels used, which emit a lot of particulates and sulphates. Some of this pollution actually has a cooling effect on the atmosphere, even though it is a horrible health hazard, and so as shipping emissions are cleaned up, this 'hidden heat' will need to be accounted for.

7. I have had an electric cargo bike for the past five years, which I use for the school run, local trips and picking up groceries; in 2021, my local London neighbourhood brought in a cargo bike loan scheme in an initiative to get more cars off the road.

8. Ken Caldeira and Ian McKay, 'Contrails: Tweaking flight altitude could be a climate win', *Nature* 593:7859 (2021), p. 341.

9. Daron Acemoglu and James A. Robinson, 'The economic impact of colonialism' in *The long Economic and Political Shadow of History: Volume I* (free ebook, CEPR Press, 2017), p. 81; I. Mitchell and A. Baker, *New Estimates of EU Agricultural Support: An 'un-common' agricultural policy*, Center for Global Development, November 2019; Nancy Birdsall, Dani Rodrik and Arvind Subramanian, 'How to help poor countries', *Foreign Affairs* 84:4 (Jul–Aug 2005), pp. 136–52.

10. Measuring economic growth is difficult because it's difficult to measure the value of all the goods and services produced by a society, and then determine whether that increased or decreased over time. One way of measuring growth is by selecting a list of products that people want, and then calculating how much of the population has access to them. This list commonly includes basic resources like clean water, sanitation and electricity. Using this metric, some countries, like Bangladesh, have seen rapid growth, whereas Chad has not. This is useful to a point but it only uses a handful of goods and services, so there is little nuance, and doesn't say anything about income – a person's options to choose goods and services. For instance, if books or tuna sandwiches were selected, would the data really tell you much about a country's economic growth? To measure the options a person's income represents we have to compare their income with the prices of the goods and services that they want. We have to look at the ratio between income and prices. This is something the data service Our World in Data does exceptionally well. Over time, for instance, the price of a book fell in Europe relative to income, particularly in the decades immediately following the invention of the printing press in the sixteenth century, which

massively increased the pace and scale of publishing productivity from the labour of individual scribes to an industrialized process. That made a book more affordable, from the price equivalent of months of wages to mere hours. And this productivity then drove further economic growth, in paper production, literacy and learning, for example.

11. Z. Hausfather, 'Absolute decoupling of Economic Growth and emissions in 32 countries', Breakthrough Institute, 6 April 2021.

12. This is partly a response to demographic shifts: Japan's construction industry is ageing quickly, with 35 per cent of all workers now fifty-five or older.

13. In 2015, China set up a Mekong River governing body called the Lancang–Mekong Cooperation Framework – 'Shared River, Shared Future' – with a broad mandate that extends far beyond river management to include cross-border cooperation in law enforcement, terrorism, tourism, agriculture, disaster response and banking. China is providing billions of dollars in loans and credit to invest in regional infrastructure, including waterways.

14. R. Barnett, 'China is building entire villages in another country's territory', *Foreign Policy*, 7 May 2021.

12. Restoration

1. A. Plumptre, 'Just 3 per cent of Earth's land ecosystems remain intact – but we can change that', *The Conversation*, 15 April 2021.

2. Eric Dinerstein, A. R. Joshi, C. Vynne, *et al.*, 'A "global safety net" to reverse biodiversity loss and stabilize Earth's climate', *Science Advances* 6:36 (2020), eabb2824.

3. A new green investment fund in the UK has been created to help restore the nation's national parks. B. Tridimas, 'Revere: New restoration financing initiative for national parks attracts corporate backing', *Business Green News*, 5 October 2021.

4. Deserts actually play a crucial role in cooling the planet since they reflect up to 30 per cent of the solar radiation that falls on them back into space and lose a further 47 per cent overnight as heat into the cloudless skies, whereas vegetated areas reflect just 10–15 per cent. Thus greening deserts could have a warming effect.

5. Alice Di Sacco, Kate A. Hardwick, David Blakesley, *et al.*, 'Ten golden rules for reforestation to optimize carbon sequestration, biodiversity recovery and livelihood benefits', *Global Change Biology* 27:7 (2021), pp. 1328–48.

6. Kelp plantations could be grown on a massive scale, and without many of the problems of terrestrial reforestation, such as fire or the need to irrigate.

7. Some scientists have proposed erecting vertical pipes through which deep waters would be drawn up to mix with surface waters. Valves would prevent the reverse flow.

8. Trish J. Lavery, Ben Roudnew, Peter Gill, *et al.*, 'Iron defecation by sperm whales stimulates carbon export in the Southern Ocean', *Proceedings of the Royal Society B: Biological Sciences* 277:1699 (2010), pp. 3527–31.

9. Ralph Chami, Thomas F. Cosimano, Connel Fullenkamp and Sena Oztosun, 'Nature's solution to climate change: A strategy to protect whales can limit greenhouse gases and global warming', *Finance & Development* 56:004 (2019).

10. Humans have dramatically altered the planet's nitrogen cycle over the last century, through industrial pollution and the use of artificial fertilizers for crop production (much of which is being washed into waterways with disastrous results), and yet nobody would suggest that we stop terrestrial fertilization.

11. Amy L. Lewis, Binoy Sarkar, Peter Wade, *et al.*, 'Effects of mineralogy, chemistry and physical properties of basalts on carbon capture potential and plant-nutrient element release via enhanced weathering', *Applied Geochemistry* (2021), e105023.

12. Phil Renforth, 'The potential of enhanced weathering in the UK', *International Journal of Greenhouse Gas Control* 10 (2012), pp. 229–43.

13. Pete Smith, Steven J. Davis, Felix Creutzig, *et al.*, 'Biophysical and economic limits to negative CO_2 emissions', *Nature Climate Change* 6:1 (2016), pp. 42–50.

14. Giulia Realmonte, Laurent Drouet, Ajay Gambhir, *et al.*, 'An inter-model assessment of the role of direct air capture in deep mitigation pathways', *Nature Communications* 10:1 (2019), pp. 1–12.

15. One method involves using potash as a solvent to dissolve the CO_2 out of the air, then builder's lime to remove the CO_2 from the potash into a limestone, which is heated in a calciner while the CO_2 is captured as a gas. To remove 10 gigatonnes of CO_2 from the air every year would require 4 million tonnes of potash, which is 1.5 times today's entire global supply of this chemical. To heat the calciner to 800°C, which is too intense for electric power alone, would require a gas furnace and gas – presumably hydrogen.

16. David P. Keller, Andrew Lenton, Emma W. Littleton, *et al.*, 'The effects of carbon dioxide removal on the carbon cycle', *Current Climate Change Reports* 4:3 (2018), pp. 250–65.

17. G. Madge, 'Temporary exceedance of 1.5°C increasingly likely in the next five years', Met Office website, 27 May 2021.

18. https://www.arcticiceproject.org

19. Yuanchao Fan, Jerry Tjiputra, Helene Muri, *et al.*, 'Solar geoengineering can alleviate climate change pressures on crop yields', *Nature Food* 2:5 (2021), pp. 373–81.

20. Phoebe L. Zarnetske, Jessica Gurevitch, Janet Franklin, *et al.*, 'Potential ecological impacts of climate intervention by reflecting sunlight to cool Earth', *Proceedings of the National Academy of Sciences* 118:15 (2021).

21. Peter Irvine, Kerry Emanuel, Jie He, *et al.*, 'Halving warming with idealized solar geoengineering moderates key climate hazards', *Nature Climate Change* 9:4 (2019), pp. 295–9.

22. Katherine Dagon and Daniel P. Schrag, 'Regional climate variability under model simulations of solar geoengineering', *Journal of Geophysical Research: Atmospheres* 122:22 (2017), pp. 12–106.

23. The SRMGI is a partnership between the UK's Royal Society, The World Academy of Sciences (TWAS) and the US Environmental Defense Fund (EDF).

24. Yixuan Zheng, Steven J. Davis, Geeta G. Persad and Ken Caldeira, 'Climate effects of aerosols reduce economic inequality', *Nature Climate Change* 10:3 (2020), pp. 220–24.

25. *Atlas of Mortality and Economic Losses from Weather, Climate and Water Extremes (1970–2019)*, World Meteorological Organization, WMO-No. 1267 (2021). Available at: https://library.wmo.int/index.php?lvl=notice_display&id=21930#.YeRUrC1oejh

Conclusion

1. I. Vazquez, 'Europe's shame: Criminalising Mediterranean search and rescue missions', *Friends of Europe*, 2 April 2019.

2. Eleanor Gordon and Henrik Larsen, 'Criminalising search and rescue activities can only lead to more deaths in the Mediterranean', LSE European Politics and Policy (EUROPP) blog, 20 November 2020.

3. Nicholas Rees, 'The climate crisis is a child rights crisis: Introducing the Children's Climate Risk Index', UNICEF website, August 2021.

4. In 1990 smartphones were nearly two decades away, there were only four TV channels in Britain, and Google, Amazon and Facebook didn't exist. A lot can happen in thirty years . . .

Index

Adger, Neil, 152

Afghanistan, 22, 139

Africa: Agenda 2063 development initiative, 87; boat-crossings to Europe from, 62–3, 134, 208; early farming in, 37; energy production in, 172, 173–4; extreme heat and drought in, xi, xii, 18, 19, 22, 110; extreme poverty in, 46, 79, 83, 122, 128, 174, 181; food production in, 154, 156–7, 158, 165, 167; genetic diversity in, 44, 45–6; !Kung peoples, 34; lack of water resources, 184, 186–7; low levels of migration to, 41–2; migration from as relatively low, 121–2; poor infrastructure and city planning, 127–8; population rise in, 30, 64, 77, 78–9, 96; rainfall due to Indian irrigation, 156–7; remittances from urban migrants, 82, 83; and restoring of planet's habitability, 191–2, 200; Transaqua Project of water diversion, 186–7; transatlantic slave trade, 41–2; transport infrastructure in, 158; urbanization in, 46, 127–8

African Union, 116

agoraphobia, 32

AI and drone technology, 165, 183

aid, development/foreign, 83–4, 158

air-conditioning/cooling, 18, 20, 22, 147, 158

airships or blimps, 177

Alaska, 13, 25, 108–9, 120, 152, 164, 184

algae, 7, 157, 162, 163, 169, 193, 195

Aliens Act (UK, 1905), 61

Alps, European, 114, 198

Amazon region, 7, 8, 13, 18, 22

Americas, 16, 30, 37, 41–2, 43, 60, 95, 127, 155

Anatolia, 35, 36, 40

Anchorage, Alaska, 152

Anderson, Benedict, 59

animals/wildlife, 155, 189–90, 191, 195; global dispersal of, 31–2; impact of fires on, 10, 12–13; impact of ice loss on, 109 *see also* livestock farming

Antarctica, 16, 106, 114, 157, 184, 195, 198, 199; ice sheet, 6–7, 8, 109, 198

Anthropocene era, xv–xvi, xviii–xix, 69, 126–7, 130–3, 155–6; four horsemen of, 10–30